# Fossil Fuels in the Arab World:
# Missed the Boat?

*Adjusting to Post-Oil Era*

# Fossil Fuels in the Arab World:
# Missed the Boat?

*Adjusting to Post-Oil Era*

## Basel Nashat Asmar

2050 consulting

First published in the United Kingdom in 2022 by
2050 Consulting
48 Imperial Hall
104-122 City Road
London EC1V 2NR
United Kingdom

A CIP for this book is available from the British Library.

ISBN 978-0-9567368-2-6

**Author's Notes**:
1. The material in the book made no use of any proprietary data owned by IHS Markit, and does not express IHS Markit's opinion. The opinions expressed in this book are the responsibility of the author.

2. The use of particular designations of countries or territories does not imply any judgement by the author as to the legal status of such countries or territories, of their authorities and institutions or of the delimitation of their boundaries.

Printed in the United Kingdom

*Cover design by Eman Faidi*

*This book is dedicated to the memory of two very special women, Alia "Lulu" Doleh, whose kind heart and selfless soul will be sadly missed by everyone who knew her and Rowida Naimat, a gentle soul who left us too soon.*

# LIST OF CONTENTS

# ACKNOWLEDGMENTS

I would like to thank my esteemed expert friends and colleagues for their insightful comments on the draft chapters for discussing and challenging key ideas in the manuscript. Special mention goes to Paul Markwell who tested some of the narrative and helped shape the book's direction and to Pritesh Patel for his support, encouragement and critical comments.

I am grateful to my friends and family in London and Amman who encouraged and supported me during the long writing process. Special thanks to Sanad and Qais, my nephews, whose smiles on video calls, throughout the lockdown, helped me while stranded at home, to write this book.

Once again, completion of this book would have not been possible without the efforts of Karen Hall, who not only edited my manuscript meticulously, but also challenged my thinking by raising shrewd questions and sharing valuable insights, which influenced the final book shape.

# ABOUT THE AUTHOR

***Basel Nashat Asmar*** is an expert in oil and gas fundamentals, costs and technology, a dynamic simulation expert with extensive computer, modelling and simulation skills. He is currently a Director with IHS Markit, based in London, UK. Dr Asmar worked previously in several roles with major engineering companies, concerned with large liquefied natural gas (LNG) regasification and liquefaction terminals, natural gas compression stations and offshore oil and natural gas production platforms. He worked as a senior process engineer and dynamic simulation specialist with CB&I (currently McDermott), Mott MacDonald, and IMEG. He was a lead consultant with 2050 Consulting Ltd and Trident Consultants Ltd. Prior to that he worked in academia as a research associate at the University of Nottingham.

Dr Asmar authored two books and published more than 50 articles in international journals, conference proceedings and newspapers. He is a Chartered Engineer, a member of the Institution of Chemical Engineers (IChemE), a senior member of the American Institute of Chemical Engineers (AIChE) and a member of the Jordanian Engineers Association. He holds a BSc in Chemical Engineering from the University of Jordan, an MSc in Process and Project Engineering and a PhD in Chemical Engineering from the University of Nottingham, and a doctorate in Geoscience from Freie Universität Berlin, Germany.

# *FOREWORD*

By
Samer Hattar, PhD[1]
*Baltimore, Maryland*
*USA*

This book is the third in a series following Dr Asmar's previous two books about Fossil Fuels in the Arab World. This book, in my opinion, is a true gem. I was tempted to spoil the plot and provide the reader with all the extensive knowledge that I gained while reading this book. Luckily, I decided against it, instead allowing the reader to enjoy the excitement of going from chapter to chapter and, interweaving all the different stories, that seem unrelated, to a coherent picture that reveals, in my mind, three incredibly important facts: First, Western democracies, which both Dr Asmar and I believe in very strongly, are in major trouble. Second, China's ascension and its relation to Russia is much stronger than our media has led us to believe. Third, the Arab world still has a chance to be a major contributor to the world, but they must act fast. As Dr Asmar expands on the following three points, one can see how climate change, democracy, fossil and alternative fuels, can all be interconnected. In his first two books, I was an outsider to these topics - I learned a lot from his two books but in this one I feel I have more of "an in" as this book is intertwined with Covid 19 and vaccine denials, two areas that fit well with my medical and scientific training.

I am delighted that my friend Basel has again asked me to provide an honest assessment of this book. As a scientist originally from the Middle East and now living and working in the medical scientific field in the USA, I was horrified reading

---

[1] The author is expressing a personal point of view here. The opinions stated in his preface should not be regarded as the official position of his employer.

this book to vividly remember the Trumpian era. (As a matter of light heartedness on a personal note, people who know me well, appreciate my fondness of the F word). That Dr Asmar described the response of Trump to the Covid-19 pandemic as a "ClusterFuck" really warmed my heart and eased some of the traumatic experiences that I felt during this Trumpian era in our history.

Before I go into the details of this book, I would like to say that this book should have a huge appeal for people of Middle Eastern origins and those interested in the Middle East. But it would be a missed opportunity if it is not read by people who are interested in Western Democracies, in the rise of terrorism, the worrying spread of nationalism, the massive increase of fake news and misinformation and the weakness of our media, due to lack of experts who can, as Dr Asmar does, connect the dots. These important points are also connected to the climate change and how the transition from Fossil Fuels to alternative sources energy is accelerating. Most interestingly for me, as an owner of an electric self-driving car (Tesla), the term MaaS (or Mobility-as-a-Service) this book will open your eyes to so many changes that are happening around you. Like the Arab world (hopefully), this book will spur you to get with the program and re-adapt your life to the new reality, or you will truly "miss the boat".

Dr Asmar provides an incredible timeline for how the last five years, (2015-2020), have accelerated change in the world that is irreversible. He continues to highlight the major setbacks that were set up by the Arab "Spring" and all the related problems that occurred because of it. Importantly, Dr Asmar correctly identifies the Arab Spring as Arab Winter, as this "Spring" has caused many conflicts that, to the time of the writing of this book, have caused major disruptions in several Arab countries, to name the most famous, Libya, Syria and Yemen. The changes that occurred in the Arab world also led to the appearance of ISIS, the most brutal form of terrorism, whose atrocities have been broadcast into our living rooms. Most impressively, Dr Asmar explains in plain words, how all the conflicts in the

Middle East Region are interconnected and how Iran and Saudi Arabia have fought their proxy wars within the Region. Reading this book, what is really saddening is to understand that the chaos of the Trumpian approach contributed to more suffering and more problems in that Region.

In my opinion, the description of the rise of MBS to power in Saudi Arabia was spectacular. In addition, the presentation of the US media coverage of this charismatic leader, which had to do a U-turn following the incredibly stupid decision to murder Khashoggi and brought his vision of modernizing Saudi Arabia to a crashing halt, was insightful and helpful. Dr Asmar explains the conflict with Qatar and the break of protocol from Trump clearly putting it in remarkable context, in a way that will make sense of the whole situation to the interested reader.

The book goes on to describe the chaotic response of the Western world to the Covid-19 pandemic. It is particularly sad when Dr Asmar compares this chaos, which cost lives, to the well-organized response of the Chinese government, which saved countless lives, admittedly completely lacking human rights and freedoms. However, the laughable response by Trump about the virus magically disappearing, treating it with light or suggesting drinking bleach, all made the situation much worse in the US.

The pandemic however, accelerated the MaaS. Dr Asmar's snapshot of the shift in technology and how that will change our lives was also a revelation. The online companies, like Zoom, that became a household name growing in power and influence incredibly fast. The economy of the service-based mobility system, where you do not have to own a car to move alone or receive your essentials, is the way of the future, especially with the generation of autonomous driving delivery and transportation.

Finally, this book delves deeply into the different kinds of alternative energies, their impacts on the environment and the costs of those transitions. It is worth considering, however, the facts that Dr Asmar lays out for us here about the limitations of alternative energies and the obstacles to their use. There is no doubt in my mind that we have to come up with alternative energy, but believing in them religiously and without considering the obstacles will not solve the severe problems we are experiencing with climate change.

This fascinating scholarly but accessible presentation, lays out the complexities of the situation, ultimately providing the reader with this significant truth - Fossil Fuels are not disappearing any time soon and the acceleration toward alternative energy sources is also not going to be stopped.

Dr Samer Hattar
*Baltimore*
*February 2022*

# INTRODUCTION

When I started writing this book early in 2020, my initial focus was to address the developments in the energy industry in the Arab world and its interactions with a rapidly changing global energy system. The global energy industry was experiencing a competition between fossil fuel and alternative sources while sustainability was trending.

However, within few weeks the focus had changed with the arrival of the Covid 19 coronavirus pandemic, when humanity went into various reactions ranging from panic mode, lockdown to denial (started March 2020).

Meanwhile, due to Russian-Saudi political rivalry, there was an unprecedented collapse in the price of oil. These unexpected and dramatic events led to a collapse in oil price which, by April 2020 created a threat to the survival for many companies. However, the crisis was short lived as the two governments reconciled and resumed the previous cooperation, via the OPEC+[1] platform, to stabilise and manage oil prices.

The Covid-19 pandemic, along with the volatility of oil price, combined to accelerate the acceptance of energy transition by the public. This will have huge implications on oil, natural gas, and future energy markets.

This book is the third in a series that spans the most significant decade in the history of the Oil Industry. The first title, *"Fossil Fuels in the Arab world: Facts and Fiction"*, examined the three types of fossil fuels: oil (petroleum), natural gas, and coal (which are in essence convertible); describing mankind's dependence on fossil fuels (particularly oil) and assessing the position of the Arab world in this picture. It addressed such topics as: the

---

[1] OPEC+ refers to OPEC members and other allied oil-producing countries.

consequences on civilisation, should the Arab World no longer provide the indispensable fuels upon which the Industrial World relies; if the fear of this might be the reason for the continuous outside political and military interference in the Arab world's affairs; and lastly, if it is justifiable for Democratic Western leaders to continue supporting autocratic and authoritarian Arab regimes to stay in power. The book examined the issue of fossil fuels in two ways, first by clarifying the terminology and language used, often misused, by talking heads, i.e. the media, politicians, scientists and the supposed "oil-experts" when discussing oil, natural gas and coal; second, by evaluating the relevant information, separating the facts from fiction - analysing the figures and scrutinising them impartially to discover the definitive quantitative data. The book concluded by examining fossil fuels in relation to alternative sources of energy, as well as from a political perspective.

The second title in the series, *"Fossil Fuels in the Arab world: Seasons Reversed "* reviewed the questions posed five years earlier, from three perspectives: market fundamentals, understanding the fossil fuel market fundamentals and the place of the Arab world within that; political influences, corruption and cultural norms in business dealings, the developing democracy and militarisation in the Arab world and their interplay with oil and natural gas were addressed and finally, public relations, perceptions or concerns, where climate change and alternative energy questions were explored in detail.

This third book picks up the story five years later examining the status of the industry and coinciding with the seismic event, that is fundamentally changing the energy industry, namely the Covid-19 pandemic.

The world is changing due to impact of concerns about climate change on government policies and shareholder and company actions. The changes will affect our way of life, with pressure being applied on both individuals and companies to curtail energy consumption and divest from fossil fuels. The current Covid-19 experience – has demonstrated the human vulnerability to the unpredicted or unexpected – have highlighted various global weaknesses and may have accelerated the energy transition.

Now THE question that must be asked, in relation to the Arab world's management of its oil and natural gas wealth is have the Arabs missed the boat? Or can they find a place in a new energy world, where the energy transition is rapidly transforming behaviour and attitudes? In a ten-year period, we have witnessed two oil price collapse periods and a cost revolution in both fossil fuel production and alternative energy sources.

In the journey to answer this question, I discovered that three main issues needed examination. Firstly, in terms of global geopolitics and macroeconomics and energy demand – what led to the split between Arab World and Rest of World? This led to considering the impact of the Covid-19 pandemic in re-setting economic development and energy demand through 2020 and 2021. Secondly, the new energy markets - looking at recalibrating energy demand post-Covid-19, the energy transition and energy financing realities, which are also redefining the Transportation sector. Thirdly, what are the options for the Arab world regarding the competitiveness of the Arab world's fossil fuels; what is being done going forward; and how to win in the future?

The text is arranged in six chapters that are structured into four generalised parts:

- Part I: Overview and Background. It consists of two chapters. The first chapter reviews briefly the events of the last five years and placing them in context with the core analysis in Part II of the book. In Chapter 2, I revisit the signposts from the previous book, and examine their status.

- Part II: Detailed Core Analysis. In this part comprehensive analysis of the energy domain globally, and in the Arab world, is presented. It consists of two detailed chapters describing oil and natural gas supply and demand and energy transition.

- Part III: Detailed Supportive Analysis. A comprehensive analysis of the transportation sector, its effects on oil and natural gas markets is presented in Chapter 5.

- Part IV: Overview. In the conclusion all the threads are woven together to form the bigger picture, identifying uncertainties and in as much as is possible, posing questions to predict what will transpire in the next couple of decades.

# Chapter 1
# *WHAT CHANGED 2015 - 2020*

In "Fossil Fuels in the Arab World: Seasons Reversed", I wrote *"In historic terms, five years does not seem to be a significant period. However, dramatic changes have occurred in this short time. Some of the events in the last five years were seismic and will continue to affect our lives for years to come."* At that time, those events seemed to occur at light speed compared to the years before it. However, I can say now, for sure, that some aspects of history repeat themselves, as the paragraph remains as true today as it was then. It goes without saying that, in the last five years (especially 2020) the world has witnessed significant changes that previously took decades to materialise, some of which were unimaginable even a few months beforehand.

The following pages highlight numerous developments that occurred in the period of interest, which have influenced current and future oil and natural gas markets and geopolitical events all around the planet.

Since the theme of this book focuses on the Arab world, I discuss, in some detail, particular events that took place there. Some of these events had global ramifications and continue to be pivotal factors in shaping global events today. In addition, I address several international events that have significant effects on the Arab world.

Three events stand out. First, the failure of the "Arab Spring", which brought hope in December 2010, but whose anticipated "Summer" never arrived, instead being followed by a harsh "Winter" where almost all political gains were brutally reversed. Second, the collapse of oil price, with all the subsequent effects on energy markets and policies.

The third significant event has been the return of the Cold War between a fragmented West and an East that is becoming more authoritarian. The unpredictable actions of Donald Trump were a major factor fuelling this war, creating havoc in international relations and decoupling the West and the East in pursuing his de-globalisation agenda. This was complicated further by the British voting for and implementing Brexit, which, for over four years, pre-occupied Europe leading to partial disengagement from the world's affairs. By the end of 2019 these three events were considered the core drivers shaping longer term trends and global future.

Tragically, lighting struck in 2020, with the emergence of the coronavirus global pandemic. The Covid-19[1] spread eclipsed all other events of the previous four years and continues to dominate the global agenda. All strategies planned and scenarios considered pre-Covid-19 were challenged as new narratives have emerged. This defining moment in global history is forcing all parties to re-consider their future assumptions, re-run their models and forecasts of how the world will be re-shaped once this pandemic is defeated, possibly leading to new global vision with energy transition, away from fossil fuels, playing a pivotal role.

In a nutshell, Trump, and Covid-19 played havoc with the world's status quo. In the Arab world, chaos reigns supreme. While Europe was paralysed by the Brexit affair. In short, international political dynamics, (especially the divisiveness of Trump and Brexit, the aggressive expansionism of Putin, the intimidating actions of China, the continuing strife across the Arab world and the chaos across South America), along with catastrophic weather events and extreme temperatures, including the increase in frequency of wildfires, storms, floods, etc., and the long-feared arrival of a pandemic, have placed the world at a crossroads.

---

[1] Scientifically known as severe acute respiratory syndrome coronavirus 2 (SARS-CoV-2), is the contagious virus causing the respiratory illness responsible for the Covid-19 pandemic.

## 1.1   Coronavirus Covid-19 Pandemic

No doubt that the most significant event that occurred in the last five years (and perhaps one of the most important thus far in human history) is the emergence and prolific spread of the highly contagious deadly coronavirus Covid-19. Since it was identified in China in December 2019, the pandemic spread rapidly worldwide, paralysed much of the world and caused the biggest economic setback since the great recession of 1929. I am certain that many books will be written on this subject in the coming years and while its significance is considerable, it cannot be the principal focus of this book.

Almost everyone knows the story of the pandemic. A once-in-a-century event, that continues to affect our lives at the time of writing. It is the first major pandemic to occur since the (mis-named) Spanish flu pandemic of 1918-20. Unlike its predecessor, this pandemic has exposed the fragility of our supposedly developed global system and laid bare the limitations of many global organisations, especially the World's Health Organisation (WHO). It also highlighted the ineptitude of many governments, some of which were caught off-guard, unprepared and without the resources to devise solutions to respond to the crisis. Many countries imposed numerous restrictions on civil liberties, including lockdowns, curfews and creating emergency powers to tackle the unprecedented situation. Alas in many countries much of their efforts were botched by the ignorance, incompetence, corruption or just plain poor judgements by politicians who sought to make trade-offs between the available capacity of their countries' health services and their economics and political positions. Thus, at the time of writing, over 114 million cases were confirmed, leading to over three million deaths.[2]

As 2020 went on, in countries previously relatively unaffected by SARS virus', the general population's knowledge of public health increased enormously. Many more people now understand the definitions of "pandemic", types of vaccines and virology jargon

---

[2] https://www.worldometers.info/coronavirus/ (as of 1st May 2021)

e.g., "infection peak", "flatten the curve", the "R-number" (the reproduction number, which is the number of people an infected person transmits disease to),[3] etc. In addition, not only has our vocabulary expanded, but so has our experience of such infection control measures as quarantine, lockdown, curfew, social distancing becoming part of the "new normal". Our working lives have transformed with working from home, becoming the new norm. Many of us are living in a new model of social interaction, with face masks becoming part of our fashion, our city centres transformed into ghost towns, with workplaces, retail, services and catering sectors forced to close to reduce the contagion within our populations.

The world strategy to control the virus relied on three pillars of parallel efforts: expanding testing for the virus, developing and distributing vaccines and developing treatments. In all three pillars, the herculean efforts paid off with treatments improving throughout the year and several vaccines being proven successful before the end of the year. The challenge of 2021 will be the manufacture, distribution and vaccination of the majority of the world's human population, educating people to ensure participation in the international vaccination programs and maintaining the infrastructure to continue these programs in response to booster vaccines or vaccines for variants that might arise in the future. This effort will be crucial not just in the developed world countries, but in all countries, with the wealthier ones ensuring that the poorer ones are not left behind. This virus and its variants are here to stay,[4] and just like the winter "flu jab", this is likely to be another annual health prophylactic treatment that may well become mandatory.

The economic and social impacts of the pandemic will be felt for

---

[3] "R", or reproduction number is a way of rating a disease's ability to spread. It is the number of people that one infected person will pass the virus on to, on average. If "R" is higher than one, then the number of cases increases exponentially, while if it is lower, the disease will eventually fade. Measles has one of the highest R numbers, with a reproduction number of 15 in populations without immunity. It can cause explosive outbreaks. Covid-19 known officially as Sars-CoV-2, has a reproduction number of about three; https://www.bbc.co.uk/news/health-52473523

[4] https://www.reuters.com/article/us-health-coronavirus-who-briefing/this-virus-may-never-go-away-who-says-idUKKBN22P2IJ?edition-redirect=uk

years. Throughout this book the virus and its impact will feature strongly while discussing most issues. In a short time we have seen that the pandemic has knocked global economic growth by several percentage points.[5] Covid-19 effects were immense as it has demolished some sectors in the economy, while allowing technology sector to thrive, e.g., online retail, online communication, and television streaming achieved forecasted ten year growth in few months period, while the share price of technology companies sky rocketed and the online communication platform Zoom, became a globally known household name.

Everyone has been affected by Covid-19 now! Many people are critical of their governments' handling of the situation. Some governments, in particular the US, politicised the situation, with Trump's hostile rhetoric calling it the "Chinese" virus spreading accusations regarding activities in the Wuhan laboratory,[6] claiming that it was "an attack on the US", demanding restitution and compensation from China. The US government even claimed that China and WHO coordinated a cover up.[7] While China initially tried to hide the facts and misinform the world, being overall less than transparent sharing the information while it attempted some damage control in the initial months of the pandemic, it later rebuked American accusations, suggesting that the virus may have been spread by American military.[8] Many argue that the pandemic might have had a very different trajectory if things had not been made worse by China's initial reactions and its restrictions on free speech.

Equally we can ask if things would have been different if the virus had emerged in a different political or more democratic

---

[5] https://www.statista.com/topics/6139/covid-19-impact-on-the-global-economy/
[6] https://www.newsweek.com/controversial-wuhan-lab-experiments-that-may-have-started-coronavirus-pandemic-1500503
[7] https://www.thedailybeast.com/white-house-asks-intel-agencies-to-find-evidence-that-china-covered-up-the-coronavirus-outbreak-says-report
[8] https://www.aljazeera.net/news/politics/2020/5/10/%D8%A7%D9%84%D8%B5%D9%8A%D9%86-%D9%88%D8%B2%D8%A7%D8%B1%D8%A9-%D8%A7%D9%84%D8%AE%D8%A7%D8%B1%D8%AC%D9%8A%D8%A9-%D8%A3%D9%85%D9%8A%D8%B1%D9%83%D8%A7-%D9%81%D9%8A%D8%B1%D9%88%D8%B3

system? The world watched with horror, as ex-President Trump created chaos in the USA, ignoring his experts' advice and spreading misinformation, e.g., advocating injecting disinfectant to get rid of the virus.[9] He created paranoia amongst the populace, stoking fear and violence across the nation, his own ignorance and stupidity being compounded by other democratically elected officials' errors. A new term to describe catastrophic incompetence on such a catastrophic scale also entered the lexicon – "clusterfuck".[10,11] Although the term dates at least as far back as the Vietnam War, as military slang for doomed decisions resulting from the toxic combination of too many high-ranking officers and too little on-the-ground information. (The "cluster" part of the word allegedly refers to officers' oak leaf cluster insignia).

The paranoid mistrust in governmental response led to many conspiracy theories gaining prominence, including "the virus is manufactured to control human race", to "no virus exists and it is all a total hoax", it is "caused by 5G telecommunications networks", and that Bill Gates is behind this effort to cull the human population, etc.[12,13,14] The WHO called the campaign to spread misinformation an "infodemic".[15] Social media, which initially helped spread massive amount of misinformation, was forced to act to curtail the spread of misinformation and fake news.[16,17] However, these efforts only slowed down, not stopped, the spread of the fake or misinformed claims around the world, leading to large numbers of the population refusing to follow health guidelines around mask wearing, social distancing an being vaccinated against Covid-19.

---

[9] https://www.bbc.com/news/world-us-canada-52407177
[10] https://qz.com/work/1225213/the-difference-between-a-snafu-a-shitshow-and-a-clusterfuck/
[11] https://www.merriam-webster.com/dictionary/clusterfuck
[12] https://www.bbc.co.uk/news/world-52224331
[13] https://www.bbc.co.uk/news/technology-52501453
[14] https://arabicpost.net/%d8%ab%d9%82%d8%a7%d9%81%d8%a9-%d8%b9%d8%a7%d9%85%d8%a9/2020/05/05/%d8%af%d8%a7%d9%8a%d9%81%d9%8a%d8%af%d8%a2%d9%8a%d9%83/
[15] https://www.who.int/news/item/23-09-2020-managing-the-covid-19-infodemic-promoting-healthy-behaviours-and-mitigating-the-harm-from-misinformation-and-disinformation
[16] https://www.bbc.co.uk/news/technology-52441202
[17] https://www.reuters.com/article/us-health-coronavirus-facebook/facebook-bans-false-claims-about-covid-19-vaccines-idUSKBN28D202

In reality, the virus exists, is highly contagious, can be fatal or leave people with serious health problems in the long term - the situation is dire.

However, in many instances, the response by some governments is exaggerated and may lead to worse consequences. The treatment should not be worse than the disease, and the medicine should not be worse than the symptoms. Examples of such exaggerated responses include the cancellation of many cancer scanning appointment leading to more severe cases, and the closure of several sectors in the economy such as the hospitality sector leading to total financial ruin to workers and businesses.

In my opinion, the way the world dealt with the pandemic is wrong, as it was always reactive, based on panic rather than planned response. The numerous lockdowns devastated economies and livelihoods of people from all walks of life. While some lives were saved from the virus, the ripple effect on those with other critical health conditions created more death from delaying treating other chronic or common diseases such as cancer, mental health, etc. Once the crisis has passed, I anticipate that we will see many unforeseen, but also difficult, long-term consequences, resulting from the mishandling of the pandemic across all societies.

A controversial question is, if it would have been better to have a more laissez faire approach, allowing the virus to pass through the population, hoping this might create a "herd immunity" where most people would survive. In my view, the moral answer is an absolute NO. We have an obligation, as humans, to avoid worsening a situation and to save life. However, I know others will disagree - and it is controversial view – and argue that one must think rationally not emotionally. Eventually, it is a Health-versus-Economy struggle.

Lower economic standards will lead to deteriorating health conditions, which becomes vicious cycle. Both Trump and Boris

Johnston's initial gut feelings were to ignore the virus and let nature take its course. Maybe they believed they were right at that time, but they were forced to change direction by intense pressure from the public and the media. It will take years of studies and reflection to reach to a semi-definitive answer. In reality, we need to understand that the efforts the world pursued for a year were futile and failed to stop the pandemic. There cannot be a sustainable solution without a vaccine, and the world cannot lockdown indefinitely. Whether we like it or not, agree or not, we need to build herd immunity sooner or later.

The consequences of the lockdowns and restrictions (on movement, economic activity, etc.) are the fiercest assault on civil liberties in decades. While some restrictions and emergency powers are justified in a public health crisis time, to minimise infections, etc., there is fear governments will not roll many of them back. This is what we saw after September 11th and yet more governments may be opportunistic in keeping some or all the authoritarian rules imposed during this crisis and retain their additional power. There is evidence that authoritarian leaders, many of whom mishandled or trivialised Covid-19 crisis, are exploiting or aggravating the response to the pandemic, placing selfish interests ahead of public good and making concentrated efforts to preserve their newly gained might.[18] Even some quasi-democratic and democratic countries are jumping on the bandwagon, with intrusive tracking tools.[19] Examples include India's intrusive Covid-19 tracing app[20] and Israel's turning surveillance tools on itself.[21] However interestingly, many citizens have been activated following the first lockdown, with some resistance being evident when governments imposed second or third lockdown to stem repeated spikes of infections. Still, many wonder what the fuss is all about, and are crying foul of the inevitable lasting economical damage.[22,23]

---

[18] https://www.theguardian.com/commentisfree/2020/apr/26/trump-to-erdogan-men-who-behave-badly-make-worst-leaders-pandemic-covid-19
[19] https://www.technologyreview.com/2020/05/07/1000961/launching-mittr-covid-tracing-tracker/
[20] https://www.bbc.co.uk/news/world-asia-india-52659520
[21] https://www.bbc.co.uk/news/world-middle-east-52579475
[22] https://time.com/5821166/gop-coronavirus-deaths-economy/
[23] https://www.theguardian.com/world/2020/mar/24/older-people-would-rather-die-than-let-covid-

In addition, the pandemic resulted in a move to isolationism and closing national borders, curtailing trade and people movement. There is a backlash to globalisation and increased nationalism, exposing vulnerability of supply chains and strengthening a desire to strengthen national borders, looking inward, returning industry back inside national borders and ignoring global interconnectivity. While Covid-19 is a global problem, apart from vaccines, there are very few global solutions to curtail this, and isolationism is certainly not the answer.

Finally, with regard to the 2020 Covid-19 pandemic, one needs to consider China, its traditions of food production and consumption, Wuhan and its laboratories, illegal pangolin and other wild animals trade, and their roles in this catastrophic outbreak.

We need to consider the lack of international coordination and cooperation, in response to the management of the outbreak all over Europe, the continent of South America, the Pan-Pacific countries and especially the US. Perhaps for some, this catastrophe has forced us to address some critical issues, not least climate change; the end of "American Century", where Trump's leadership has left America "weaker, and sicker, and poorer" and how we can learn from this crisis so that we can do better next time.

It is important to realise that Covid-19 pandemic outweighed and overwhelmed all other world events, diminishing their importance and reducing, or even trivialising, their impacts, e.g., the cancellation of the World Cup and the Olympics. The remainder of the chapter discusses the events that happened prior to Covid-19 pandemic, which without Covid-19 would have been more prominent.

---

19-lockdown-harm-us-economy-texas-official-dan-patrick

## 1.2 Energy Transition's Unstoppable Momentum

Energy transition is a significant change in the structure of energy systems, which is currently underway and has gained an unstoppable momentum. It relates to the replacement of the, currently dominant, fossil fuels with more environmentally friendly energy alternatives in order to combat climate change and reduce greenhouse gases (GHG) emissions. One cannot talk about changes in the last five years without stressing the importance of this transition and its consequential impacts upon the Arab world. Most of the world's superpowers, international governments from every continent, multinational organisations and industries are pushing to reach their lower GHG emissions targets. They are formulating policies and changes to achieve a net zero objective. Pressure groups, environmental activists, scientists and local communities are also applying pressure to push these changes through.

While it is crucial to introduce the subject here, equally, reducing energy transition to merely another change in the last five years, underestimates the important implications, across a variety of issues, on both the Arab world and the other international communities. In order to explore the issue sufficiently, "Energy Transition" is covered in more detail in Section 3.6.

Chapter 5 focuses on the transport revolution, detailing how this new way of life will affect the future of oil and natural gas and, as a result the financial, social and political consequences on the petrostates, in particular those within the Arab world.

The Covid-19 pandemic with its resultant lockdowns have demonstrated possible new ways of living and amplified the efforts towards net zero. In my view, this is what now drives the future of oil and natural gas. It has the most significant impact on both the Arab world, as a fossil fuel major producing region and the developed world, as a major consumer.

Several scenarios are being envisaged for the future of energy usage, with the IEA, EIA, OPEC and many oil and natural gas companies rushing to devise appropriate new flexible policies and strategies. These are discussed in detail in the coming chapters.

## 1.3   The Arab World in Flux

As with my last book, I will not be discussing the Arab-Israeli conflict in any detail. Despite the fact that it is not only a core issue, but is also linked to many other Middle Eastern challenges – the issue is beyond the scope of this book. In my opinion, what is really telling about how stagnant the situation has become, is the muted reaction by the Arab world to the Trump's government recognising Jerusalem as the capital of Israel, moving its embassy to Jerusalem in 2017.

The political instability in the Arab world (triggered by the events following the Arab Spring revolutions in the early 2010s) appeared to be subsiding until, in 2019, brutal regimes emerging from the Arab Winter took control. Unfortunately, due to increasing economic hardships and erratic American international policies, a second wave of Arab unrest sparked in 2019, threatening to destabilise the region again.

In the following sections, I discuss the major events that have shaped the Arab world in the last five years. These events, when seen in a wider context, allow us to understand the complex picture that forms the new geopolitical map of today's Middle East.

The Arab-Israeli conflict effectively turned into a narrower Palestinian-Israeli conflict in 2020, when four Arab countries (UAE, Bahrain, Sudan, and Morocco) normalised their relationships with Israel, it looks like a few more are about to follow suit. This narrow conflict continues to drag on, showing no signs of being resolved. It has become a regular news item, with its flares of violent incidents occurring frequently in a confined geographical region - many cannot even agree how to refer to it[24] - and it seems to be accepted now as an ongoing humanitarian tragedy.

Since Hamas took control of the Gaza Strip in 2006 – practically splitting the Palestinian Authority, there has been unrelenting conflict between Israel and the self-declared Hamas government. Israel continues to impose a siege on this Palestinian entity. The conflict has turned into a desperate saga, where it sporadically flares into what have become full-blown wars, with the latest major episode fought in the summer of 2014 and skirmishes have continued since then. It seems that the Palestinian-Israeli conflict has gone on for so long that people in the region have become de-sensitised, almost resigned to it. Despite occasional media outcries, hollow statements of regret, condemnation and calls for restraint from various public figures, we watch these wars and the descent of the lives of the occupants of the Gaza Strip, into unliveable hopeless conditions, without taking any effective steps to stop them. A much hyped "Deal of the Century" unveiled by the Trump administration in 2020, failed to materialise and the situation remains unresolved. It seems that no-one has the belief or the will to find solutions to end this terrible conflict.

### *1.3.1 The Arab World's Search for a New Season*

After the Arab Spring, with all the hopes it brought to the young people failed and so many aspirations shattered. The period between 2010 and 2015 could be described as an era of dashed hopes and the turning of euphoric optimism into dark pessimism. The Arab Winter that followed, brought a darkened mood, a deep pessimism and a bleak political future.

In all the Arab countries where the democratic transition failed, authoritarian rule returned with vengeance. Vicious dictators seized power and re-established harsher police states. They manipulated the innovative methods used by the original

---

[24] An indication of how complex the situation is, the region has at least eight different names and is referred to as the Holy Land, the Occupied Territories, the Palestinian Territories, the Palestinian Authority, the West Bank and Gaza Strip, Judea and Samaria, etc.

protesters during the Arab Spring, i.e. digital technology and social media, to effectively distribute the propaganda of the new regimes. Many of these regimes restricted the internet and censored free speech. They enacted laws to punish those using internet-based free expression and created electronic armies (referred to as flies) of bots to spread fake news, alongside their propaganda.

The change of foreign policy by Trump's administration took many Arab governments by surprise. His American administration threw out the familiar political rule books and his actions at home and abroad became predictably unpredictable. Many unmovable old principles of diplomacy, communication and international cooperation were scrapped, and previously fixed positions became moveable.

Despite the chaotic nature of the Trump Administration, many Arab governments followed his lead, instigating internal policies that disregarded human rights and restricted media freedoms. They were encouraged by his government's attitude turning a blind eye of many suppressive policies, and thus felt emboldened pursuing increasingly illiberal policies. His endearing nickname of the Egyptian president as his "favourite dictator" was received pleasantly!

In the Arab monarchies which were affected less by the Arab Spring, some governments' wise actions managed expectations (except Bahrain), and life continued relatively smoothly. Some of the reforms enacted as a result of the Arab Spring were quietly rolled back. In the last five years, there were uneventful royal successions in Saudi Arabia and Oman. King Salman succeeded King Abdullah in Saudi Arabia (see Section 2.1 for developments in Saudi Arabia) and Sultan Haitham succeeded Sultan Qaboos in Oman, after a fifty-year rule.

However, in many of the Arab republics, the apparent calm that followed the Arab Winter appeared to be superficial, with discontent continuing to simmer, while in others tensions often boiled over and the fires of the Arab Spring continued to burn.

One change that has established itself is that the tenure of Arab leaders is shortening. Prior to the Arab Spring, most rulers would have been in power for an average of 15 years in the region, the duration of those terms in power are noticeably changing. See Figure 1.1.

Initially the Gulf Cooperation Council (GCC) countries' reaction to the Arab Spring started in a unified manner, however, their policies soon diverged, as the Arab Winter went on. Inevitably each country had its own agenda and struggled to maintain a unified front. Instead, individual countries pursued different policies, supporting opposing parties in Syria, Libya, Iraq, and Yemen.

The rift between the GCC countries widened enormously, culminating in 2017 when Saudi Arabia, UAE and Bahrain cut all relationships with Qatar and imposed a crippling blockade to bring it to submission. They were joined by their ally Egypt. Together they demanded a series of conditions that were impossible for Qatar to accept, including the closure of Al Jazeera tv station. Surprisingly Qatar endured. During this period of conflict, the US administration was divided, with Trump siding with Saudi Arabia, while his Secretaries of State and Defence siding with Qatar. As time progressed, the US realised the importance of Qatar to its policy objectives and changed its position, asking the countries of the GCC to reconcile. The situation remained tense for several years despite the many diplomatic attempts trying to bridge the differences. The situation thawed in January 2021 with an agreement restoring ties between the countries.

The regionally influenced internal conflicts that resulted from the Arab Spring had different outcomes. Some concluded, while other fires are still burning. To date three internal conflicts are raging, in the form of vicious civil wars in Yemen, Syria and Libya. These wars have created unprecedented refugee crises,

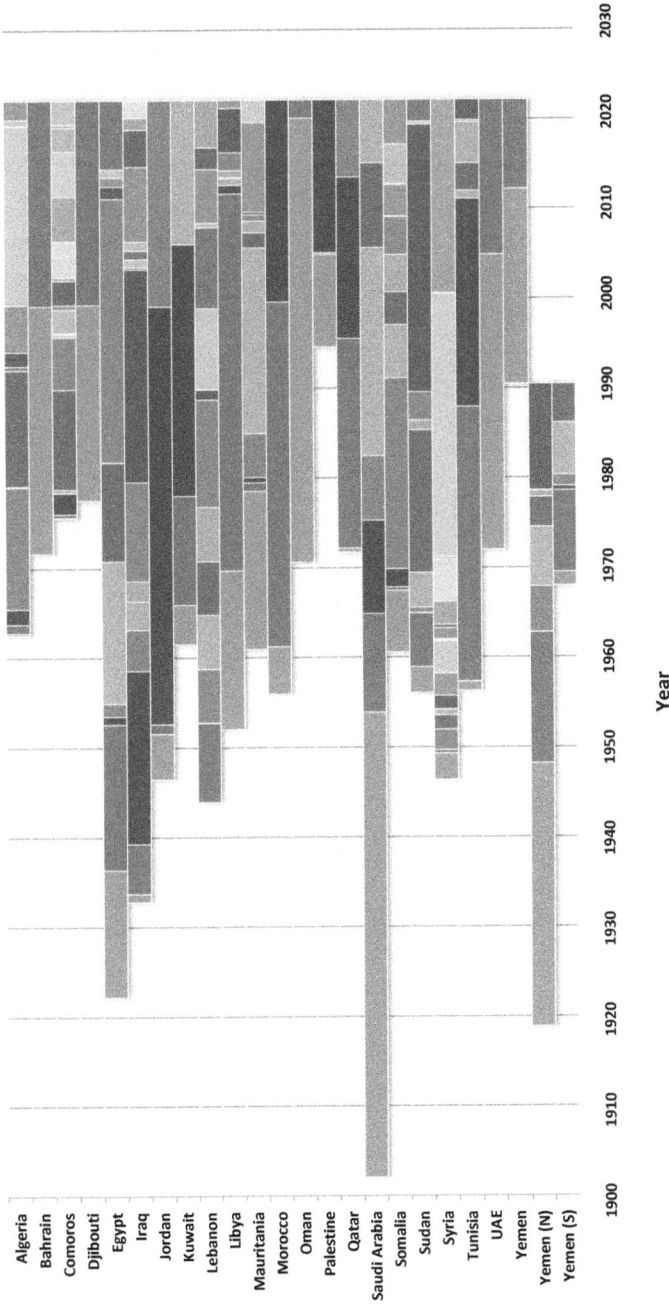

*Figure 1.1: Length of reign of control of Arab leaders since independence*

Source: calculated by the author

with a record number of internally displaced people. Figure 1.2 shows the decline in the population of Syria since the start of the war.

The three countries have turned into proxy theatres of war between the regional powers, with Saudi Arabia and Iran squaring up to each other, subsequently involving world and regional superpowers (Russia, the USA, France, Israel and Turkey) who intervened to support their respective allies. The conflicts in Syria and Yemen have fuelled sectarian conflict between Sunnis and Shias that is redrawing the ethnic, religious and political map of the Arab world.

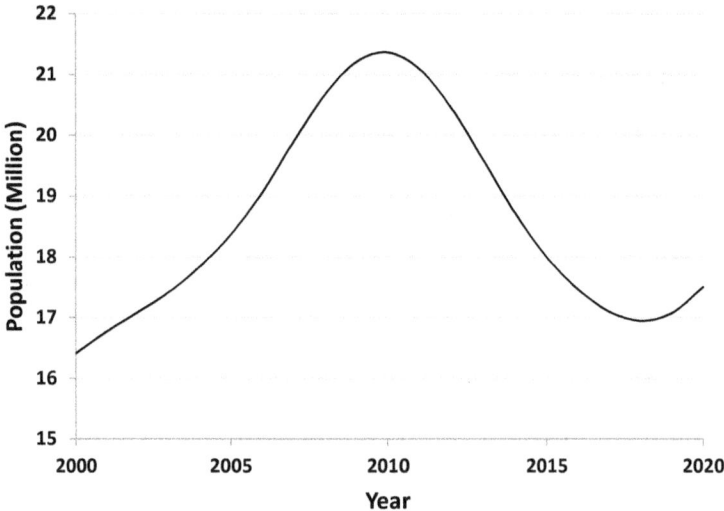

*Figure 1.2: Syria's population (2000 - 2020)*
Source: Statistica (https://www.statista.com/statistics/1067100/population-syria-historical/)

The first closed conflict came to a swift conclusion in Bahrain, where the Sunni minority government prevailed. At that time, GCC rulers supported the Bahraini government in supressing its Shia uprising. The three main GCC countries, Saudi Arabia, UAE and Qatar were united in resisting increased Iranian influence on their doorsteps. But as nothing is certain in politics, Bahrain did not automatically support Qatar and instead is a vocal critic of its policies at the moment.

In Egypt political suppression also prevailed, The Arab Spring revolutionaries were dealt the devastating blow of a military coup, which successfully overthrew the democratically elected president and brought about the return of harsh military rule. President Sisi rules with an iron fist. His two predecessors died: Mubarak of natural causes and Mursi (the only democratically elected leader in Egyptian history) in suspicious circumstances in prison.

The Syrian civil war continues. It appears that President Assad has effectively destroyed his country, buts otherwise prevailed. The war has turned Syria effectively into a Russian protectorate. No doubt that Assad and Russia were the main winners. The war is coming into a slow conclusion, but it has turned into a type of conflict reminiscent of the "Cold War".

In the last five years, the warring factions succeeded in flushing out ISIS from their controlled territories in Syria. While the Western-supported Kurdish rebels made significant territorial gains at the expense of ISIS, the hostility from Turkey towards them, enabled the Assad regime to reinstate its authority in many regions. His reach is increasing slowly, with rebels from all factions reduced to shrinking enclaves.

The massive refugee crisis, which peaked in 2015, has now subsided with cooperation from Turkey stemming the flow of humanity to Europe. However political differences with the EU, its history of human rights abuses and increasing tensions with Cyprus are delaying the acceptance of Turkey's admission to the EU. However, there is a risk that Turkey could restart that flow whenever its leadership wants, if strategic conflicts continue to occur between itself and the EU.

The political instability in Iraq continues. ISIS initially controlled roughly 30% of Iraq's land in the west and northwest; the Kurdistan Regional Government (KRG) took advantage of the chaotic situation, and - with American and European support - claimed several disputed territories from the central government, contributing to the disintegration of the Iraqi state. The tide turned in 2017 with the defeat of ISIS and the subsequent reclaiming of most disputed areas from the KRG by the Iraqi central government. However, rife corruption and political, instability caused unrest to flare up again in 2019, forcing the government to resign in March 2020. Maintaining a stable government remains difficult as regional and international powers all continue to meddle. The situation in Iraq remains totally chaotic, the fractured parliament cannot enact policies or enable significant change and the country is virtually bankrupt.

In Libya, following the overthrow of Colonel Gaddafi, a short-lived stable government was formed. However, rivalry amongst the former allies who had overthrown the old regime, jostled to control the county and internal conflicts flared up. The country entered a vicious cycle of civil wars. It became another proxy war involving Egypt, Saudi Arabia, France, Turkey and the UAE, with the latter emerging as the chief influence, supporting a potential dictator and paying Sudanese militias. Local allegiances continue to shift and the internationally recognised government control has, once again, been reduced to pockets in the north-western part of the country where the capital was besieged. However, with assistance from Turkey, the recognised government turned the tide. Sadly, the country remains divided, with all efforts for resolution failing thus far. Unless progress is made and an effective government formed, Libya's disintegration continues.

The brutal conflict in Yemen, led by Saudi Arabia's Muhammad bin Salman (MBS) continues. This proxy war, involving Iran, UAE and Saudi Arabia – with militia imported from Sudan and Pakistan has ostensibly brought Yemen and its peoples back to the Middle Ages.

In 2015 former President Saleh supported the Iranian-backed Houthi rebels in northern Yemen in order to regain power, but after appearing to betray them, he was killed by Houthi rebels who consolidated their grip on most of northern Yemen. Initially close allies, the Unity Pact between Saudi Arabia and the UAE appears to have crumbled, with the two countries pursuing different agendas and supporting different factions. There are indications that the UAE has aspirations to control parts of the South Yemen and Socotra Island, while the Saudis have set their sight at Mahra governorate and gaining direct access to the Indian Ocean. At the moment the situation has turned into open civil war, with a total collapse of law and order in the country.

Furthermore, the Houthis have extended their reach attacking deep into Saudi Arabia and the UAE. They gained massive media attention following a successful devastating attack on oil installations in eastern Saudi Arabia in September 2019 knocking out over 50% of Saudi oil production temporarily.[25]

Sudan and Algeria, two countries initially not affected by the Arab Spring, started a wave of protests in 2019, often referred to as the Second Arab Spring. The protestors seem to have learned from the original Arab Spring and adapted their techniques accordingly. Both protests succeeded. In Sudan, President Bashir (who was under indictment by the International Criminal Court (ICC)) was deposed and arrested after thirty years of military rule. In Algeria the twenty-year rule of President Bouteflika, who at the latter years was a zombie president, ended. The new regime succeeded in sending the former president's corrupt brother and former de-facto ruler, to prison.

Finally, in Lebanon, the populations' patience reached a breaking point in 2019, following the decision by the nearly bankrupt, corrupt government to impose a new tax on internet telecommunications. This triggered what was dubbed the "WhatsApp Revolution", forcing the government to step down. The unrest in Lebanon continues, despite the change of the prime minister, with protestors now demanding a total overhaul of the

---

[25] https://www.reuters.com/article/us-saudi-aramco-fire/attacks-on-saudi-oil-facilities-knock-out-half-the-kingdoms-supply-idUSKCN1VZ01N

sectarian political system that was initiated and installed as part of a deal to end the 1975-1992 Civil War.

The catastrophic explosion of a huge amount of an abandoned ammonium nitrate, stored at Beirut Port, exposed the dire state of the Lebanese government and illustrated to the entire world the level of collapse the country has reached. The explosion caused over 200 deaths, 6500 injuries, 300,000 people made homeless and US$15 billion in property damage.[26] The involvement of rival regional powers, including Saudi Arabia, Iran, Israel and even the former colonial power France, have not helped to improve the situation.

So far Tunisia remains the only successful country to have benefitted from the Arab Spring. It is the sole Arab country that made the transition to a fragile democracy. While the situation there is increasingly tense, two peaceful presidential successions following free elections, seem to have cemented the young democratic system.

In the preceding paragraphs I have given a brief overview of the political turmoil that continues to trouble the Arab world. There is a wide array of published material available, which is an abundant source of information for those readers who want to understand the history of the Region but which is outside the scope of this text.

The socio-political turbulence in the Arab world continues to cause significant disruption in oil and natural gas production, which has had considerable effects on several countries around the region. Figure 1.3 shows the oil production in selected countries and the overall production in the Arab world, while Figure 1.4 shows Egypt's natural gas production and exports. To interpret the graphs, one needs to consider that:

- Oil production initially collapsed in Libya, and its recovery in 2014 was considered a significant factor,

---

[26] https://en.wikipedia.org/wiki/2020_Beirut_explosion

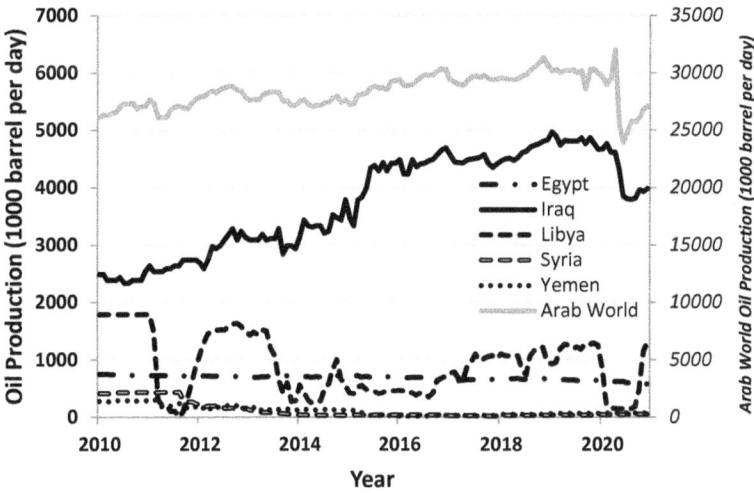

*(a) Quantitative production*

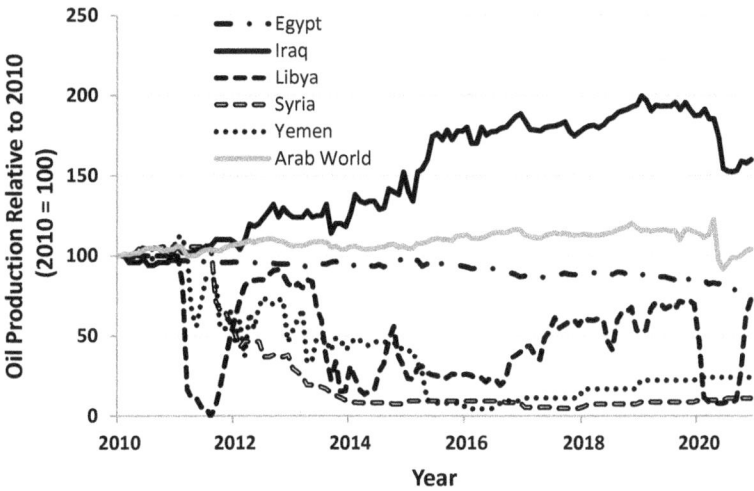

*(b) Production change relative to 2010*

*Figure 1.3: Crude oil and condensate production in selected Arab countries (2010 - 2020)*

Source: EIA
Note 1: Total Arab world production is on the right hand axis in Figure (a).

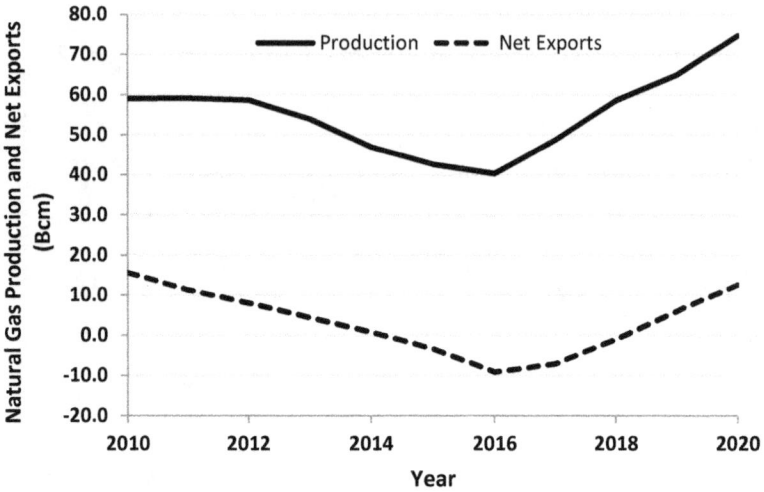

*Figure 1.4: Egypt's natural gas production and net exports (2010 - 2020)*
Source: Cedigaz; EGAS, BP Statistical Review

contributing to the oil price collapse then. Libyan oil production has fluctuated since then, but in 2020 fell to a record low, due to continued internal fighting, before recovering later in the year.

- Oil production collapsed totally in Syria, where the majority of the producing fields first fell into ISIS hands, then to the Kurdish rebels. The limited current production now is mostly from Kurdish controlled areas.

- Some oil producing fields changed hands several times in Iraq, where the KRG took control from the federal government, who subsequently took them back. In addition, several fields fell under ISIS control, were then taken by the KRG and then the federal government. Despite these events, Iraq's overall oil and natural gas production grew substantially in the last five years.

- Yemen's oil production collapsed totally due to the ongoing civil war. However, although its natural gas production was initially maintained and liquefied natural gas (LNG) exports continued,[27] operations were subsequently halted. Several attempts to restart the LNG terminal stalled due to the ongoing security situation.[28,29]

- Security issues contributed to reduction of natural gas production and the cessation of gas exports from Egypt. Between 2015 and 2018 Egypt has become net natural gas importer. However, the discovery and rapid development of the giant Zohr natural gas field, changed its fortunes and repositioned it as exporter once again. See Section 1.3.3 for more details.

It is interesting to note that despite all this disruption, the overall Arab world production increased over the same period. This is due to the policy adopted by Saudi Arabia and its allies to continue to increase their production and thus flood the oil market to defend their market share. The consequence of this policy lead to the collapse of the oil price in 2020 as discussed in Section 1.5 that follows.

### 1.3.2   Miscalculations of the Kurds

In the last five years, the Kurdish goal of establishing an internationally recognised state has not materialised. Although the Kurds number almost 40 million people[30], they are one of the largest ethnic groups in the world, outside India, without an independent homeland.

The problem is that their homeland is strategically located between four powerful competing Middle Eastern countries (Turkey; Iran; Iraq and Syria) with further pockets in the Caucasus region. These four countries are often engaged in conflicts and disputes, but there is one issue upon which there is consensus – opposition to an independent Kurdistan. These four countries often manipulate and/or supress the Kurds, obstructing

---

[27] http://en.terra.com/news/news/yemen_lng_says_pipe_blast_to_cut_exports_by_four_cargoes/act471680

[28] https://oilprice.com/Energy/Crude-Oil/Can-Yemens-Oil-Industry-Make-A-Comeback.html

[29] https://www.offshore-energy.biz/total-hopeful-of-yemen-lng-production-restart/

[30] Including a diaspora of over two million people, mainly in Europe.

them from achieving their goal of a separate State. Despite many attempts, the realisation of an internationally recognised Kurdish state remains unlikely.

Several episodes in recent history brought the Kurds close to achieving their independence goal. However, each one ended in disappointment. At the end of World War I, following the collapse of the Ottoman Empire. the Treaty of Sèvres in 1920 promised the Kurds an independent homeland in territories currently located in southeastern Turkey.[31] However, after the Turkish War of Independence, the Treaty of Lausanne in 1923[32] superseded the previous articles of Treaty of Sèvres, and the Kurdish people remained stateless. Since then, there have been numerous Kurdish uprisings against the rule of Turkey, Iran, Iraq and Syria. The four surrounding countries used these uprisings to their advantage, fermenting conflict between the factions, supporting certain Kurdish factions as instruments against neighbouring countries and repeatedly dividing them against each other. These cycles have continued up to the present day.

Relative to other Kurdish groups, the Iraqi Kurds appear to have made the most progress towards the goal of statehood. In Iraq, following intense unrest in 1960s, the Kurdish Autonomous region was established in 1970. However, the implementation of autonomy proved impossible and the bloody conflict in Northern Iraq resumed.

The 1990 Iraqi invasion of Kuwait, and subsequent American-led Gulf War of 1990-91, triggered a Kurdish uprising, which, with the military help of Western powers, who imposed a no-fly zone, succeeded in expelling the Iraqi government forces. Following

---

[31] The Treaty of Sèvres was signed between the victorious Allied powers (United Kingdom, France, Italy, Greece, and Armenia) and representatives of the government of the defeated Ottoman Empire. The treaty abolished the Ottoman Empire and forced Turkey to relinquish all its Arab territories in the Middle East and North Africa, and the division of the proper Turkish territory into influence zones controlled by the signatories.

[32] The Treaty of Lausanne was signed by representatives of Turkey as successor to the Ottoman Empire, and the victories allies (Britain, France, Italy, Japan, Greece, Romania, and the Kingdom of Serbs, Croats, and Slovenes) concluding World War I. The treaty recognized the boundaries of the modern state of Turkey and imposed no controls over its finances or armed forces. It guaranteed free shipping through the Turkish straits.

the Anglo-American invasion of the country in 2003 overthrowing Saddam Hussein and his Baathist regime, for the first time, Iraqi Kurdistan had real control of its affairs. Iraqi Kurdistan finally gained official recognition as "sovereign".

The Iraqi constitution of 2005 identified Iraq as an independent federal state, with Iraqi Kurdistan deemed a federal entity. It combines the provinces of Erbil, Dohuk and Sulaymaniyah.[33] However, there is growing divergence between the Kurdish Regional Government (KRG) and the central Iraqi government. Since 2005, the Iraqi government in Baghdad has recognized the KRG, exercising its authority governing, and controlling all internal and, most, external affairs of Iraqi Kurdistan autonomous region. However, there remains a dispute between the two governments regarding who has direct authority over parts of the provinces of Kirkuk, Nineveh, Saladin and Diyala.

In the aftermath of the Gulf War of 2003, the Region became even more unstable. The poorly reasoned disbanding the Iraqi army weakened the Iraqi central government, which in turn, splintered the country into three major ethnic and religious factions. The Sunni Arabs in the North and West became marginalised, providing ideal conditions for radicalisation, which enabled extremist Islamic groups (e.g. ISIS) to prosper.

Following the ISIS capture of Mosul in 2014 and the attempted invasion of Kurdistan, the KRG took advantage of the unrest and occupied many of the disputed areas, including the city of Kirkuk with its vast oil reserves. This action emboldened the KRG and in 2017, against advice from all their international allies, the Kurds went ahead with a controversial independence referendum in their attempt to legitimise their independence. This act backfired spectacularly when the Iraqi government retaliated, ousting them from Kirkuk and most other disputed areas, while their allies stood aside and refused to assist them. Current negotiations have failed to resolve the issue of Joint Control in the region.

---

[33] Halabja was carved out of Sulaymaniyah and proclaimed a fourth governorate by the KRG in 2014, but this move had not been ratified by the Iraqi federal government.

The killing of Qassem Soleimani[34] in 2020 created many difficulties for the KRG. It is expected that, as the Iraqi parliament and government are requesting, American troops will eventually leave Iraq. Once that occurs, the KRG is vulnerable without American support, if it were to be attacked.

Despite the setback losing Kirkuk, the KRG continues to behave as a de facto independent country. It has a government which pursues independent foreign and economic policies, it controls its borders and issues its own visas. In addition, new laws being introduced in the region are progressively diverging from federal laws. Effectively the region is politically divided into two sub-entities or administrations with the Kurdistan Democratic Party (KDP) controlling the Western Erbil and Dohuk Governorates, while the Patriotic Union of Kurdistan (PUK) controls the Eastern Sulaymaniyah and Halabjah Governorates.

Owing to decades of under-exploration, oil and gas reserve numbers in the KRG are inaccurate. Oil reserve estimates in the KRG vary considerably,[35] with projected numbers ranging from 4 billion barrels (according to the International Energy Agency (IEA)), to between 45 and 60 billion barrels (according to the KRG officials). The latter figure has not been independently verified, and possibly includes unproved resources.[36] In addition the latter, much larger estimate, includes reserves from the disputed areas, so whatever the amount, ownership is unclear. Other unsubstantiated reports state even larger estimates. The Sulaimani Provincial Council claims that the Garmiyan block, in their province, has 30 billion barrels of oil alone.[37] These figures seem to have been accepted locally and are quoted by the DOR Organization for Kurdistan Oil and Gas Information. In 2015 they published a report about the Kurdistan Region's oil and gas

---

[34] Qasem Soleimani was an Iranian major general in the Islamic Revolutionary Guard Corps (IRGC) and a commander of its Quds Force. He was considered the second most powerful person in Iran. He was assassinated in a targeted American drone strike on 3rd January 2020 in Baghdad.

[35] https://www.eia.gov/beta/international/analysis_includes/countries_long/Iraq/iraq.pdf

[36] Refer to Section 3.1 for clarification regarding fossil fuels technical terminology.

[37] http://rudaw.net/english/kurdistan/220720151

sector, announcing that it is the world's eighth country (in terms of its oil and natural gas reserves), claiming that it will be last place in the world to run out of oil.[38] Dr. Ghalib Mohammed Ali, head of the Oil and Gas Committee of the Sulaimaniya Provincial Council in his report, claims that Kurdistan Region, as part of Iraq, has 50 billion barrels of proved reserves and 80 billion barrels of unproved reserves. In addition, he claims that it has eight to ten trillion cubic meters of natural gas reserves which, if this were true, would indeed make it one of the largest untapped supply resources on the planet.

Since its de facto independence, oil and natural gas exports accounted for the majority of the KRG's revenue. The vast oil and natural gas reserves and their production enabled the regional government to bolster its independent status while pursuing their own separate oil and natural gas policy. These actions have been opposed by the Iraqi federal government.

Following the de facto separation from Iraq, oil and natural gas production in the KRG has fluctuated due to the disputes with the federal government. However, despite this, in comparison to the rest of Iraq, the KRG has benefited from this period of relative stability. The liberal oil and natural gas laws have attracted investment from foreign companies. The KRG has recently expanded its oil and natural gas infrastructure, significantly developing production fields. Over the last five years, it expanded its export capabilities, which besides exporting oil by road to Turkey and Iran, also built export pipelines, linking it to Turkey, thus enabling its oil to be exported via Ceyhan port. These policies benefited the KRG's government and its economy where oil production and exports increased steadily. See Figure 1.5.

This unusual independent policy has led to numerous disputes between international oil companies and the federal Iraqi government, who threaten companies with sanctions if they invest in projects in the KRG without the consent of the federal government. However, this deterrent has proven ineffective, as

---

[38] http://ekurd.net/iraqi-kurdistan-has-worlds-8th-oil-and-gas-reserves-2015-03-30

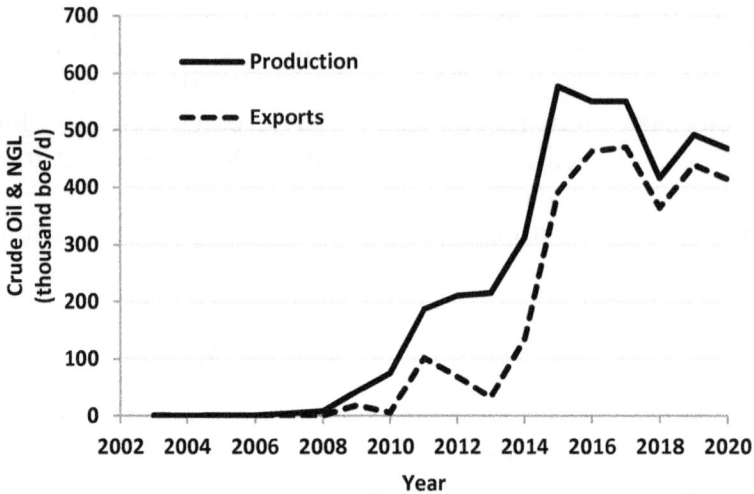

*Figure 1.5: Crude oil and natural gas liquids production and exports in KRG*

Source: http://mnr.krg.org/index.php/en; MEES, argusmedia.com; Rudaw.net

oil companies continue to invest in the KRG, rather than federally controlled Iraqi resources. In fact, some of the major international oil companies have prioritised their investments in the KRG.

Meanwhile, across the border in Syria, Kurdish rebels benefitted from the Syrian civil war, establishing an autonomous administration, commonly known as Rojava. The defeat of ISIS initially bolstered the Kurdish goal of independence with American and Israeli support. However, the prospect of establishing a successful Kurdish entity caused Turkey to interfere militarily to derail the initiative. Furthermore, Kurdish aspirations for and Independent State are being jeopardised by the strained relationships between the Kurds in Rojava and the KRG. In the last two years American support to Rojava has weakened, sadly, once again, the Kurds are being abandoned by their allies.

### 1.3.3   Eastern Mediterranean Growing Gas Discoveries

In the last two decades, the discovery of natural gas in the Eastern

Mediterranean region has transformed energy politics there. While earlier discoveries transformed natural gas markets in Egypt and Israel, additional discoveries in the last few years have bolstered the Eastern Mediterranean region's reputation as a major new hydrocarbon province and transformed it into a promising area for oil and natural gas exploration.

Prior to this millennium, Egypt was the sole natural gas exporter in the region. It actively supplied its neighbours via a network of pipelines and sold LNG on the international market. However, Egypt overstretched its resources and became a natural gas importer in 2015 (see Figure 1.4). In August 2015, the Italian oil giant ENI announced a massive natural gas discovery in Eastern Mediterranean, the Zohr field, the largest discovery in the region, off the shores of Egypt. The commercial reserves were estimated to exceed 30 Tcf. This discovery along with other smaller discoveries changed everything and Egypt reversed its fortunes by accelerating the development of the field. It is currently in a position to supply its growing demand and provide natural gas exports to fulfil its previous export contracts regionally and internationally.

Beyond Egypt, between 1999 and 2000, several small natural gas discoveries were made in Palestinian and Israeli territorial waters. These were followed by significant discoveries - the Tamar natural gas field in 2009, followed shortly afterwards by Leviathan in 2010, with a combined volume exceeding 26 Tcf. The timing of the discoveries was ideal since it coincided with the rapid decline in natural gas supplies to Israel from Egypt, whose domestic demand was rapidly increasing. This also forced the Egyptian government to halt its natural gas exports to Jordan,[39] as well as stopping its LNG exports. These discoveries allowed Israel to become self-sufficient and turn the country into natural gas exporter.[40,41,42]

---

[39] Exports to Jordan were severely affected by the deteriorating security situation in Sinai and the gas pipeline linking Egypt and Jordan was subjected to numerous sabotage attacks.
[40] https://www.reuters.com/article/us-israel-natgas-leviathan-idUSKBN1YZ0H9
[41] https://www.spglobal.com/platts/en/market-insights/videos/market-movers-europe/011821-nord-stream-oil-energy-forum-iea-sustainability-week-french-nuclear

Elsewhere in the Eastern Mediterranean, significant natural gas discoveries were made offshore of Cyprus and drilling activities also started offshore of Lebanon.

The discovery of Cypriot natural gas is an issue of conflict between Turkey, Greece and the Greek Cypriot government. Turkey's protégé, Northern Cyprus, claims rights to these natural gas resources, while Turkey is pursuing its own exploration activities in the region with no success so far.[43] The dispute is thorny, as several countries' territorial claims overlap. In addition, Turkey's involvement in Libya has complicated matters further with yet additional contradictory territorial claims, which is creating tensions between Turkey and the EU, with France meddling strongly in the affair.

Other potential resources may yet be discovered in the territorial waters of Egypt, Israel, Lebanon, Syria and possibly Greece, Turkey and Libya. While the discoveries to date are significant, they fall short of the 122 Tcf estimated to exist by the US Geological Survey.[44] The civil war in Syria, delayed natural gas exploration there, although recent agreements with Russian companies reinstated this item to the agenda.

At the moment, Egypt is pursuing an ambitious plan to become a regional natural gas hub, exporting Israeli and future Cypriot natural gas. At the same time, it is involved in an expensive (maybe not needed)[45] plan to construct an offshore natural gas pipeline to transport Israeli and Cypriot natural gas resources to Europe.[46] Although the pipeline plans are progressing on paper,

---

[42] https://www.offshore-mag.com/production/article/14074908/leviathan-gas-flowing-to-jordan
[43] Despite recording no success in the Eastern Mediterranean waters Turkey announced a major discovery exceeding 14 Tcf of natural gas in the Black Sea (https://oilprice.com/Latest-Energy-News/World-News/Turkey-Expands-Gas-Drilling-In-The-Black-Sea.html;
https://www.upstreamonline.com/exploration/turkey-increases-tuna-1-gas-catch-to-14-3-trillion-cubic-feet/2-1-895387)
[44] The Middle East conflict over natural gas, Ahmed Al- Bassosy, Regional Center for Strategic Studies' monthly publication of the "State of the Region"
[45] https://euobserver.com/green-deal/149929
[46] https://www.nsenergybusiness.com/projects/eastern-mediterranean-pipeline-project/

they are facing problems and may prove hard to materialise.[47] Turkey is firmly opposing this effort and prefers export routes solutions crossing Turkey, which are more feasible. However, with high tensions between Turkey, its neighbours and the EU, the situation remains fluid. But who knows? Maybe relations between Egypt and Turkey will improve following the end of the blockade of Qatar? – But, after a seventy year cycle of wars, peace, progress and destruction that does not look like it will end anytime soon, where alliances often change – anything could happen.

### 1.3.4 *The Reversing Fortunes of South Sudan and Ethiopia*

The division of Sudan in 2011 into South Sudan and Sudan (effectively North Sudan) was a bitter divorce following long rebellions and civil wars spanning over 38 years. It was a stark demonstration of the failure of policy of the then Sudanese President Bashir who, despite losing a third of his country, clung to power without shame or remorse. The failure was exacerbated by leaving many issues unresolved, leading to enduring tensions and several conflicts between the two countries. Incredibly, the final borders between the two states have not been fully demarcated, and each government continues to support rebel groups operating in its neighbour's territory, thus reducing any chance of stability.

Soon after the formation of South Sudan, the rival rebel groups who had formed a coalition government, reignited their old tribal animosities and their new government collapsed. The tensions between the factions descended into military confrontations, engulfing the country in bloody ethnic conflicts. These conflicts escalated, dragging the country into a drawn-out civil war from December 2013 to February 2020. A fragile peace has held since a Peace Agreement was signed in 2018.

As a result of the long conflict, the country's economy has been devastated and all hopes of building a progressive democratic

---

[47] https://www.dw.com/en/eastmed-gas-pipeline-flowing-full-of-troubling-questions/a-51871424

state have stalled. According to the Council on Foreign Relations "Global Conflict Tracker", this conflict displaced over two million human beings and killed an estimated 400,000 people since 2013. South Sudan has, for several years, been listed as the world's most fragile state,[48,49] and is placed near the bottom, on the majority of the world's political and economic indicators.

While South Sudan retained control of approximately 70% of the oil reserves and oil production,[50] as all the oil export pipelines crossed through North Sudanese territory, South Sudan was forced to remain dependent on those export pipelines. This situation allows the North Sudanese government leverage over policies in South Sudan.

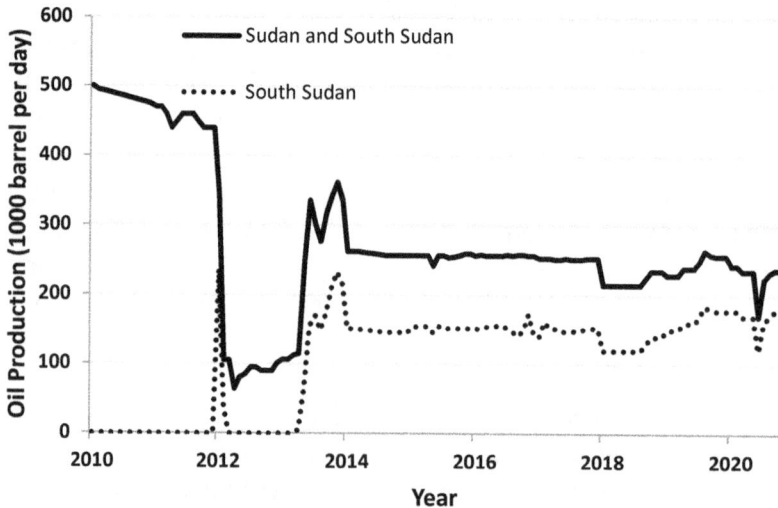

*Figure 1.6: Crude oil and natural gas liquids production in Sudan and South Sudan*

Source: EIA; http://www.voanews.com/content/oil-conflict-revenue-investment-production-instability-sudan-globaldata/2859593.html

---

[48] http://www.businessinsider.com/the-worlds-25-most-failed-states-20147?op=1#ixzz3eRX8ZKyW

[49] See Section 4.5.

[50] http://www.eia.gov/beta/international/analysis_includes/countries_long/Sudan_and_South_Sudan/sudan.pdf

The disagreements between the two Sudanese states slowed oil production after independence, and the long civil war in South Sudan contributed to further decline in their oil industry over the years. The latest cessation of military activities boasted production, but it remained below earlier levels as illustrated in Figure 1.6. Interestingly, the oil production operations in this region are dominated by Chinese and Malaysian companies. Those governments are pursuing policies to secure their oil supplies, increase their influence in Africa and their goal is to replace the influence of Western companies. As has been the case for centuries, the colonial chessboard that is the Continent of Africa, continues to be where East and West vie for control of valuable resources, to the detriment of the native populations and their environment.

In contrast, Ethiopia's political and economic prospects have improved significantly in the last few years. The ascendance to power in 2018 of Prime Minister Abiy Ahmed, an Oromo,[51] following several years of unrest and protests, gave the second most populous country in Africa a new direction.

Hailing from a mixed religious household, he promoted religious harmony and reconciliation in a country divided by religion. He initiated political reforms in the country and freed political prisoners and opposition figures. He also ended the long conflict with Eritrea. His policies earned him the Nobel Peace Prize in 2019.

This newly found political stability, successful policies, coupled with years of high economic growth has ignited hope and brought relative prosperity. The growing confidence of the Ethiopian government allowed the government to pursue significant infrastructure projects, e.g., with the Grand Ethiopian Renaissance-Dam, which started in 2011 but was facing political

---

[51] The Oromo people are a Cushitic ethnic group and nation native to Ethiopia who speak the Oromo language. They are the largest ethnic group in Ethiopia accounting for around a third of the population.

and funding obstacles. Unfortunately, Egypt, who opposes the dam construction, has tried hard to prevent the construction of the dam, which it considers a great threat. Sudan vacillated about the dam, as it, alongside other African countries in the Nile Basin, consider the inequitable rights to the use of Nile water between the countries of the Nile Basin, signed by British colonial powers, to be unfair.

The Egyptian government and media consider the project to be an existential threat to Egypt, since the filling and management of the water of the dam will deprive Egypt of its main water source and supply.

Despite American diplomatic efforts, tensions remain high between Egypt and Ethiopia. Many Arab media outlets accuse Israel of meddling, encouraging Ethiopia to harm Egypt. Unless agreement is reached through diplomatic channels, the risk of military escalation cannot be ruled out.

Late in 2020 the situation in Ethiopia deteriorated, with an uprising and military conflict in Tigray regional state. This conflict is threatening many of the achievements gained in the last few years.

### 1.3.5 Iran Nuclear Deal: Slow Death, Disintegration, or Revival?

The uncertainty surrounding the future of the Iran nuclear deal continues to cast a shadow on the geopolitics of the region. The big question is what will happen with Iran, following the events regarding the Iran nuclear affair?

The situation began in 2003, when the International Atomic Energy Agency (IAEA) revealed that Iran, in breach of its international legal obligations, had not declared a militarily capable nuclear program. The discovery led to a well-publicised dispute with a stick and carrot approach trying to offer economic incentives in exchange for giving up the nuclear ambitions, or else face sanctions. There were numerous rounds of lengthy negotiations, as well as tough international sanctions to try to remove what was perceived as the "Iranian nuclear threat".

When the conflict intensified in 2012, further tough sanctions were imposed on Iran to curb its oil exports and thus deprive the country of its major revenue stream. As a consequence, Iranian oil production and exports collapsed (see Figure 1.7).

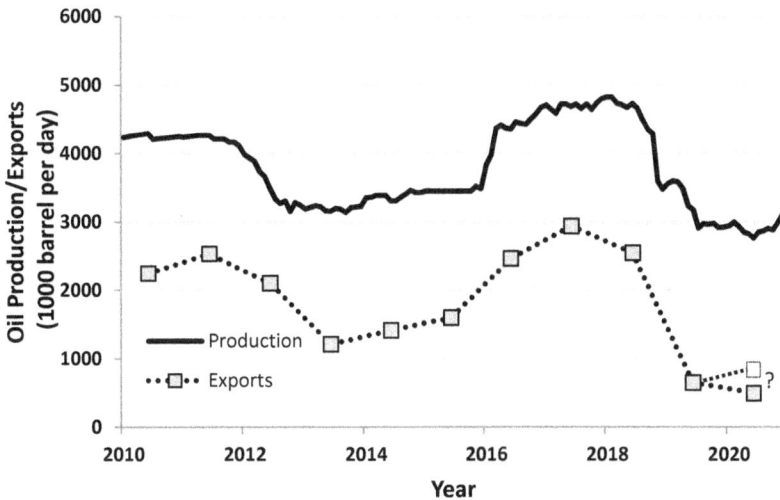

*Figure 1.7: Crude oil and natural gas liquids production and exports in Iran*

Source: EIA; OPEC, https://www.everycrsreport.com/reports/R46213.html,
https://www.ceicdata.com/en/indicator/iran/crude-oil-exports
Note 1: Includes condensate
Note 2: Export numbers are yearly averages.

After almost a decade of tough negotiations, in order to curb Iranian nuclear ambitions, the Joint Comprehensive Plan of Action (JCPOA) was signed by Iran and the P5+1 countries.[52] Iran, in exchange for the lifting Western sanctions, agreed to an intrusive inspection regime of its nuclear facilities restrictions to curb its nuclear ambitions, alongside the gradual normalisation in international relations. The deal was met with fierce opposition

---

[52] The five permanent members of the United Nations Security Council (China, France, Russia, United Kingdom, United States), plus Germany and the European Union.

from Israel and several Arab countries, who remain highly suspicious of Iran's true intentions and the risks they pose to the region's stability. At the time, it was hoped that implementing the agreement would be a game changer in the Middle East and with the global oil and gas markets.

Lifting the sanctions translated to the resumption of Iranian oil exports. Despite scepticism from many analysts, Iran succeeded in ramping up its production by up to 1 million barrel/day within six months of its return to market,[53] (see Figure 1.7) and successfully regained part of its market share.

The election of Donald Trump in 2016 led to the future of the agreement becoming endangered. During his election campaign, Trump publicly described the agreement as "very bad" and vowed to either withdraw from it or cancel it altogether. He carried out this threat and the US declared its withdrawal from the agreement in May 2018. The US reinstated, then gradually tightened, economic sanctions on Iran. Consequently, Iran oil production and exports collapsed[54] (see Figure 1.7 illustrating the plummeting in the numbers), depriving the Iranian government of its major revenue source. President Trump vowed to exert "maximum pressure" to force the Iranian government to concede to American demands and renegotiate the deal.

The sanctions pushed the Iranian economy into a deep recession and the value of the Iranian rial halved. They continue to have major impacts on Iran leading to living costs rising dramatically, creating substantial hardships to its economy and population. This suffering helped trigger widespread protests in November 2019, that were brutally quashed by the authorities.

Following US withdrawal from the agreement, Iran subsequently withdrew its compliance and began flouting the terms of the deal, strengthening its nuclear capabilities. Repeated tension between

---

[53] http://www.bloomberg.com/news/articles/2016-01-03/iran-won-t-harm-oil-market-with-post-sanctions-production-boost
[54] https://www.bbc.co.uk/news/world-middle-east-48119109

the US and Iran endures, with several crises occurring in the last two years - including the US assassination of a top Iranian general, and Iran retaliating by shelling a US base in Iraq. Despite all the rhetoric, both countries have managed to avoid direct military confrontation, albeit they are engaged in numerous proxy wars elsewhere, as discussed in Section 1.3.1.

Meanwhile as the proxy conflicts are intensifying, Iran has commenced a resettling project for three strategic islands, disputed by the UAE, to tighten its grip on the islands.[55] The tension is expected to drag on, especially after the normalisation of relations between the UAE and Israel.

Following the 2020 American election, the Biden administration has taken office, now the deal looks like it is back on the cards. Thus, after several years of Iran being suppressed, conditions there may soon improve. So, while optimism turned into grave pessimism in 2017, in 2021 pessimism may recede and optimism may return again.

If the deal is revived then, in the following inevitable power struggle between Iran and Saudi Arabia, Iran once again will be established as a dominant power in the Middle East, once again a major player influencing the oil market.

## 1.4    Cold War 2.0

After the fall of Berlin Wall in 1989, the world naively believed that the Cold War has finally come to an end. How wrong we were. In recent years the growing hostilities between the USA and China and Russia have demonstrated that this was merely a pause and that the Cold War perhaps never ended. We can clearly see the emergence of "Cold War 2.0" in the deterioration in relationships between Russia, China and the West. Unlike the 20th century, the epicentre of economic activity has moved East,

---

[55] https://www.aljazeera.net/news/politics/2020/5/1/%D8%A5%D9%8A%D8%B1%D8%A7%D9%86-%D8%A7%D9%84%D8%AC%D8%B2%D8%B1-%D8%A7%D9%84%D8%AA%D9%88%D8%B7%D9%8A%D9%86-%D8%A7%D9%84%D8%A5%D9%85%D8%A7%D8%B1%D8%A7%D8%AA-%D9%87%D8%B1%D9%85%D8%B2

as evidenced in the energy demand that is consistently growing in Asia.

During the Trump administration, the Americans single-handedly dismantled the foundations of the international world order that has been in place since the end of the First Cold War.

The West is no longer united. There is a growing rift between the USA and Europe, as well as between Europe and Britain. During Trump's term of office, the value of the North Atlantic Treaty Organization (NATO) was challenged. Its existence was under threat as Trump, often using his fat thumbs firing uncontrolled tweets, destroyed all cornerstones of American policy with Europe and appeared to favour Russia, despite the hostility towards Russia by the rest of his country.

President Trump pulled out of numerous international agreements and organisations including the Trans-Pacific Partnership, the Paris Agreement on climate change, the World Health Organisation (WHO), the United Nations Human Rights Council, the Intermediate-Range Nuclear Forces Treaty as well as many other international treaties and organisations. The Biden Administration is already reversing many of those ill-conceived actions by the most controversial president in American history.

While these facts provide relevant context events in this book and no doubt some of my personal opinions will colour my views, I will not be discussing them here - there are many others far better qualified then I to take on that epic subject. However, I will briefly address two main issues in the following sections.

### 1.4.1   *Russia Is Guilty by Default*

In the last five years, Russia's political stance has changed significantly. It has displayed more aggressive, antagonistic policies abroad, while implementing deeply conservative policies and curtailing freedoms locally. Within the Russian State, it is evident that there has been a deliberate curtailment of civil liberties, an attack on Russian democratic institutions, a pursuit of ultra-nationalistic policies while also targeting of ethnic and sexual minorities. The Russian authorities appear to be on a mission to "Make Russia Great Again".

The revival of oil prices, post 2016, gave the Russian government the fiscal muscle it needed to reassert its positioning on the global stage. Its oil revenue soared until 2019. Fortunately, Russia had grown wiser after the oil price collapse episode in 2014 and had established a fund to save all excess oil revenue, above its planned budgets, against future difficulties, whatever they might be. This foresight helped the government weather the oil price collapse resulting from the oil supply war with Saudi Arabia and the demand destruction related to the pandemic.

Moreover, the country has enhanced its capabilities in cyber weaponry, where it excelled, and has for many years been utilising its capabilities as an effective strategic tool in pursuing its political objectives. In many cases it proved more effective and lucrative than its traditional weapons trade. In fact, following a massive growth of 127% in military spending between 2010 and 2015, Russia increased military spending by a further 10% in local currency terms between 2015 and 2020.[56]

It appears the West once again failed to learn from historical precedents how to deal with Russia and Putin's implicit goal of restoring the Soviet Union in a new guise as a modern Russian Empire. While Europe and the US continue to prevaricate, the Russian government has ramped up its rhetoric against the West and become more involved with its neighbours' affairs.[57]

The Russians belief is that when it comes to interfering in the affairs of former Soviet republics, they can get away with murder. In addition to annexing the Crimea in 2014, Russia now controls two puppet states that they created in Ukraine. They are

---

[56] http://www.sipri.org/research/armaments/milex/milex_database
[57] http://www.dailymail.co.uk/news/article-2961643/Russia-tensions-trigger-war-Britain-s-NATO-commander-warns-Soviet-style-tactics-pose-existential-threat-being.html

also involved in the political affairs of Georgia and Moldova, where they continue to control unrecognized puppet states in both countries, threatening their territorial integrities. While some sanctions have been imposed against Russia, overall, despite the media coverage, they have been short lived and ineffective.

Emboldened by its success in the Crimea and Ukraine, Russia continued to be aggressive. On the world stage, it has gained confidence and made decisive steps involving itself in the affairs of Middle East, first in Syria, then in Libya. Using old techniques from the Cold War and the new technology of the internet, it interfered in US elections, the referendum on Brexit in the UK and the subsequent election there. There is significant evidence that Russia has developed sophisticated electronic and cyber weaponry that have been very effective in those two national elections (who knows what other clandestine activities have gone on in those or other countries). Russia has certainly mastered these techniques to affect change, influence opinions, sow discord and mistrust, changing expected outcomes, e.g., election results.

The ineffective muted responses by Trump to Russian policies and actions, his appearance to be in league with Russia, cannot be explained rationally. It certainly seems to be at odds with all previous American strategies and policies. The relationship between Trump and Putin suggest a more suspicious, sinister hidden reason.

Russian support turned the tide in the Syrian civil war, ensuring a military victory for Assad's government, against all the odds. It also ensured the survival of Russia's last military base in the Mediterranean.[58] After five years of direct Russian military involvement and support for President Assad's government, their intervention in Syria cemented the position as a major power in the Middle East.

---

[58] Some may interpret this as another chapter in the emerging Cold War between Russia and the West. Others may challenge this interpretation of events and offer other reasons behind Russia's motives or agenda.

It appears that the Russian government, under Vladimir Putin, has already begun to pursue its expansionist policies. The actual decision has been taken for sure, and after the Crimea and Ukraine, it is only a matter of time the next incursion happens. What will happen when Russia tests the resolve of the West further? It is a not a question of "if" but "when", Russia will invade one of its neighbours.

Prime targets are Ukraine, again, or the Baltic countries. In any instance, Putin can once again use a tested German tactic used in World War II – protecting the Russian speaking minorities which he used before in Ukraine - as the pretext to justify further military action.

Another region of potential conflict could be the Arctic, where Russia and four Western countries have overlapping territorial claims. The strategic positioning of the region is increasing since climate change can make the shorter trade route navigable.

Besides its military strength, Russia has strengthened its position by utilising its dominant role, supplying natural gas to Europe. It supplies in excess of one third of EU needs[59] and a dominant share of other European countries needs also. Despite some European countries attempts to find alternative supplies in the medium term, most are currently heavily dependent on Russian natural gas for their energy needs in the shorter term. In addition, Russia continues to expand its pipeline infrastructure to entangle European consumers more. In the last few years, it expanded exports to Europe via Turkey.[60] It is on the verge of completing the controversial Nord Stream 2 pipeline that links Russia and Germany, bypassing Poland and Ukraine.[61] These moves have found fierce opposition from the US and provoked increasingly tougher sanctions from the Americans.[62]

---

[59] https://www.statista.com/statistics/1021735/share-russian-gas-imports-eu/
[60] https://www.offshore-energy.biz/gazprom-starts-delivering-turkstream-gas-to-bosnia-serbia/
[61] https://www.ogj.com/pipelines-transportation/pipelines/article/14189195/nord-stream-2-pipelay-resumes
[62] https://www.argusmedia.com/en/news/2173670-us-congress-authorizes-new-nord-stream-2-

In response, the Russian government has opened alternative markets for exporting its natural gas supplies. Despite doubts from analysts,[63] Russia has strengthened its natural gas ties with China operating a planned pipeline, planning additional export pipelines to increase capacity and exporting LNG.[64,65] The two countries are forging closer alliances with mutually advantageous solutions to the hostile European and US policies against them.

Lastly, Russia surprised many by building a strategic alliance with Saudi Arabia to manage oil price. Its cooperation via the Opec+ framework, allowed it to play an increasingly important role in manipulating oil markets. Early in 2020 Russia deliberately triggered an oil price war so they could get rid of US shale and try to teach the Saudis a lesson. But they miscalculated the level of drop and had to relent, resuming their cooperation with Saudi Arabia.

The historic rivalry between the Russians and America is entrenched to a degree that may yet take decades to resolve. In an incident reminiscent of the space race of 1950s.[66,67], we saw, once again, how fiercely the competition between these old enemies still goes. In the race to successfully produce a Covid-19 vaccine, Russia rushed the development of its version, claiming it won the vaccine development race. It symbolically and provocatively named its vaccine Sputnik V.

### *1.4.2   China's New Positioning as the Second Superpower*

In 2015 the economic miracle of China's previous two-decades, started showing signs of weakness and appeared to be running

---

sanctions

[63] https://www.csis.org/analysis/russia-china-gas-deal-and-redeal

[64] https://www.naturalgasworld.com/china-and-power-of-siberia-ii-ngw-magazine-79532

[65] https://www.argusmedia.com/en/news/2165620-china-extends-russian-gas-import-pipeline-update

[66] https://www.washingtonpost.com/world/russia-unveils-coronavirus-vaccine-claiming-victory-in-global-race-before-final-testing-is-complete/2020/08/11/792f8a54-d813-11ea-a788-2ce86ce81129_story.html

[67] https://www.thelancet.com/journals/laninf/article/PIIS1473-3099(20)30709-X/fulltext

out of steam. The unprecedented long period of economic growth, averaging over 10% annually,[68] transformed the country beyond recognition. Its nominal GDP climbed to second place globally, after the United States. However, in purchasing power parity, GDP China has already beaten the United States to the top spot.

In the last decade, China's reputation for manufacturing quality has significantly improved in the eyes of the world's consumers, from the cheap imitator to a sophisticated inventor.[69] It is no longer the "world's factory" of shoddy goods and cheap labour. Its products are no longer seen as nasty, single-use-only goods, but rather, are held in high regard in many sectors, especially technology-related, where they are competing to be market leaders.

Chinese companies are growing and appearing more prominently in global lists, with considerably higher rankings than in the past. In 2020, the Fortune 500 list of companies included 124[70] from China, an increase from 98[71] in 2015.[72] Currently, numerous Chinese companies are expanding aggressively across a range of sectors, i.e. electronics, white goods, telecommunications, energy and banking. Some Chinese brands are becoming household names, the global leaders in their markets, e.g. Lenovo, Haier, Huawei, Xiaomi, Oppo, Alibaba, Tencent, Baidu, ByteDance,[73] Didi Chuxing, Sinopec, CNPC, CNOOC, China Industrial & Commercial Bank of China (ICBC), China Construction Bank Corporation (CCB), and Agricultural Bank of China (ABC).

Domestically, the standard of living has improved dramatically. A larger proportion of the population made the leap into middle

---

[68] https://data.worldbank.org/indicator/NY.GDP.MKTP.KD.ZG?locations=CN
[69] The Economist, July 11th 2015, 62, citing MIT Sloan Management Review & McKinsey Global Institute.
[70] https://fortune.com/global500/
[71] Fortune, November 1st 2015, p8
[72] Note that certain Chinese media outlet already brag that Chinese companies outnumber the US companies (for example, https://news.cgtn.com/news/2019-07-22/China-tops-U-S-in-number-of-companies-in-Fortune-Global-500-IxATitjK6s/index.html), however the 129 number quoted includes 10 Taiwanese companies.
[73] Owner of Tik Tok

class, fuelling internal consumer market growth, which has transformed the economy from an export-based economy into a service and consumer spending-based economy. This could have ramifications on China's imports of certain raw materials and energy products, with its ripples affecting the global markets. However, the country's economic trajectory has also exposed fragilities in the economic model going forward, amongst them are ageing population, a shrinking labour force and increased labour costs. It is noteworthy that the economy was, in part, artificially stimulated by pursuing unnecessary infrastructure mega projects, which inflated property prices and an anticipated debt crisis in the property market, like that of 2008, is looming.

While the economic transformation into a controlled capitalist model was seen as resounding success by many, the transition from communist dictatorship to a pseudo-democracy, pursuing liberal social policies, started to reverse following the election of President Xi Jinping in 2013. He tightened his grip on power, created a personality cult, removed time limits on presidency terms, re-empowering the security services to take harsh actions when dealing with opposition and dissent.

As a result of Trump's anti-Chinese policies from 2016-2020, President Xi was able to capitalise his popular support, ignore Western pressure and pursue his nationalist agenda. He imposed strict censorship on information, suppressed liberties and continued perpetrating human rights violations, targeting ethnic and religious minorities, particularly the Muslim Uighurs in Xinjiang and cracking down on Hong Kong democracy activist protestors.

On the global stage, the Chinese government forged ahead with its economic dominance plan. Its "Belt and Road Initiative", marketed as a global development strategy, involves infrastructure development and investments in nearly seventy countries and international organisations, including three NATO members (Turkey, Greece and Italy), as well as Israel, one of the US closest allies.

President Trump declared his dissatisfaction with China's influence in the American economy and viewed the growing trade deficit between the two countries as an unfair attack on US industry. He considered that China took advantage of unbalanced international treaties to benefit, at the expense of the US. He started a crusade to eliminate, what he saw as, the imbalance in the relationship between the two countries, starting a policy of decoupling the US from China.[74] Since 2018, the US and China have been engaged in a trade war, resulting in many tariffs imposed on trade between the two countries; the imposition of restrictions and sanctions on many Chinese companies, especially high-tech, along with other provocative actions. This has led to some damaging responses by China, to counter the effects of those hostile American policies. Despite many rounds of negotiations, interim truces, a resolution to the stand-off appears to be elusive at the time of writing, with the new Biden administration continuing to stand firm against China.

In fairness, while the US has legitimate grievances, it also spread fake news to justify and support its position. While China previously lagged behind other nations in technological research and development, nowadays, they have overtaken the US. So, it is inaccurate to portray China as an intellectual property thief anymore.[75] Supremacy in new technology is one of the fiercest battlegrounds in the ongoing trade war, where control of the next generation of internet protocols and telecommunication standards will provide a major advantage.

The telecommunications conflict with China led to the US banning its businesses dealing with ZTE (the partially state-owned technology company that specializes in telecommunication). Although that ban was subsequently lifted, the US then blacklisted Huawei, the world leader in 5G technology. It banned businesses from using Huawei technology, claiming that using their hardware/software potentially threatened

---

[74] https://www.ft.com/content/42aa2664-1c12-11ea-9186-7348c2f183af
[75] https://www.ft.com/content/26903a94-3617-11ea-ac3c-f68c10993b04

national security, the infrastructure that runs the country and potentially endangers lives in an entirely new form of cyberattacks. It has pressured its allies to do likewise, with the UK, Australia and others following suit.

This ongoing technological war is splintering the internet, inadvertently helping China in its efforts to create its own internet ecosystem. The Great Internet Wall of China, known officially as the Golden Shield Project, continues to strengthen, causing the censored Chinese intranet to become even more isolated.

This hostile relationship, the repeated confrontations regarding hacking, diplomatic incidents and repeated accusations of espionage may trigger an all-out cyber war between the US and China.

The Chinese government is growing more confident in pursuing its "Sinofication" policy internally towards minorities, especially in Xinjiang and Tibet. With its almost total control of the media, it is effectively fuelling nationalistic fervour by promoting nostalgic stories of past greatness from ancient Chinese history. Similarly, the Chinese government is showing no hesitation in imposing its grip in Hong Kong, moving swiftly to crush protests in 2019 and working to remove the democratic rule of law there. Other government directed narratives are broadcast by the state-controlled media to whip up patriotic fervour against real or imagined threats from "enemies". Its diplomatic policies have reignited a long-standing dispute (since 2010) with Japan over an uninhabited group of Islands in the East China Sea; it portrays Taiwan as a rebellious part of China that needs to be brought to heel; it broadcasts propaganda against the US with sustained criticism (even more so after the Trump years). All the while, it continues stoking diplomatic hostilities, offering strategic protection to the North Korean regime, as its erratic dictator acts out on the world stage.

China is asserting its authority more strongly and flexing its muscles in the South China Sea. It grows more aggressive in enforcing the nine-dash (recently ten-dash) line, claimed as its maritime borders, which covers most of the sea, threatening navigational freedom in a vital sea route to global trade. The claim overlaps the "Exclusive Economic Zone" claims of Brunei, Indonesia, Malaysia, the Philippines, Taiwan and Vietnam. This claim has created territorial disputes with each of these nations, giving China strategic control of important shipping lanes, which affects a quarter of the world's trade. In order to cement its claims to disputed islands and atolls, the Chinese have created physical facts in the disputed water, where they have been pushing a man-made island-building program.[76,77] establishing permanent airbases to counter the might of the American naval aircraft carriers. China has embarked on a massive expansion of its military power in the last five years, modernising its navy and air force,[78] adding aircraft carriers, building new land and ballistic missiles, modernising its nuclear power facilities and embarking on a space militarisation programme.[79,80] They are determinedly working to displace the USA as the dominant military power of the world.

At present, there is an escalating race of space militarisation between the USA, Russia and China. Each of these countries are building up controversial new capabilities to wage war in space, developing missiles capable of destroying satellites, independent global communications, GPS navigation systems, meteorological satellites, military monitoring and surveillance systems.[81,82] In 2019 China became the first nation to land on the dark side of the moon.[83] China is strengthening its space research and it is already planning moon and asteroid mining. Armed super-powers in space will threaten not only each other in space but also possibly

---

[76] http://www.bbc.co.uk/news/magazine-35031313

[77] https://www.defensenews.com/opinion/commentary/2020/04/17/chinas-island-fortifications-are-a-challenge-to-international-norms/

[78] https://chinapower.csis.org/china-naval-modernization/

[79] https://www.eastasiaforum.org/2020/12/16/chinas-military-modernisation/

[80] https://economictimes.indiatimes.com/news/defence/china-attempting-to-militarise-space-as-it-seeks-to-modernise-its-military-power/articleshow/77851406.cms

[81] Are we on the Cusp of War in Space?, Scientific American, October 2015, p.11-14.

[82] https://www.theatlantic.com/technology/archive/2020/07/space-warfare-unregulated/614059/

[83] https://www.nationalgeographic.com/science/2019/01/china-change-4-historic-landing-moon-far-side-explained/

threaten the inhabitants of the Planet Earth below. This will of course, escalate international tensions and create challenges.

China's actions are often openly aggressive, e.g. threatening and chasing out the oil exploration and drill-ships belonging to other parties in the disputed waters, (Vietnam[84] and Malaysia[85,86]). There have also been several standoffs with the US navy.[87]

Similarly, following years of expanding activities in Antarctica, rumours are getting louder that China intends to make a territorial claim there. China is showing greater tendencies to interfere in the Arctic, where the promise of a shorter navigable trade route is of strategic importance to future Chinese trade. China is investing heavily to exert influence in Danish-administered Greenland (Kalaallit Nunaat), attempting to establish a foothold in the region. Trump's attempt to purchase Greenland was met by total fury by local and Danish politicians who declared that "the island not for sale". If we think that Polar exploration involves working in inhospitable regions, China is also pursuing deep-sea mining,[88] moon mining[89] and mining in outer space.[90]

Progressive Chinese government leaders have achieved huge progress with their goal of becoming THE world leader. Patriotism and nationalism remain the foundation of Chinese policy and the "China Dream", the return to greatness, remains the ultimate goal.[91] Implementing this policy has been successful, transforming the country into a world-class economic and military power. It has also lifted hundreds of millions of people out of poverty, creating the largest "middle class" consumer population in the world. An important step towards achieving the ultimate dream is the "Made in China 2025" strategic plan,

---

[84] https://www.energyvoice.com/oilandgas/asia/258490/beijing-oil-china-vietnam/
[85] https://www.oedigital.com/news/478402-petronas-drillship-exits-south-china-sea-after-standoff
[86] https://www.energyvoice.com/oilandgas/asia/241105/beijing-triumphs-in-south-china-sea-oil-spat/
[87] https://news.usni.org/2020/05/08/u-s-sends-warships-on-patrol-near-south-china-sea-standoff
[88] https://chinadialogueocean.net/10891-china-deep-sea-exploration-comra/
[89] https://www.voanews.com/science-health/china-joins-race-mine-moon-resources
[90] https://www.mining-technology.com/news/origin-space-launch-mining-robot-space/
[91] https://www.bbc.com/news/world-asia-china-22726375

announced in 2015, to upgrade the manufacturing capabilities of Chinese industries to a more independent technology-intensive powerhouse, that is less reliant of foreign supply chains. It aims to move away from producing cheap low-quality goods, instead producing higher-value products and services, i.e. pharmaceutical, automotive, semiconductors, IT, aerospace, automation industries and robotics. The recent tariffs and sanctions imposed by the US, from the Chinese point of view, vindicated this policy and added to its urgency.

The ultimate goal envisages a Chinese-led 'New World Order' where the Chinese government dominates a regional bank, world trade and global state-owned companies. It is coordinating with Russia to move away from using the US dollar in international trade.[92] As part of its' strategy, China is promoting the idea of an alternative reserve currency to replace the dollar.[93] The irresponsible fiscal policies pursued by Trump in 2020, including quantitative easing and strong stimulus policies, have destabilised the US dollar, severely shaking the international financial system.[94] Recently China took further actions, announcing their electronic currency as a new initiative to make this a reality.[95,96]

However important all the events above, are, they have been overshadowed by the Covid-19 pandemic and its aftermath, which will determine the position of China on the world stage for years to come.

The pre-Covid-19 era can be defined by the growing tensions between China and the US. These tensions escalated when Trump assumed office in 2016, his leadership and disruptive diplomatic approach, including making wild (not often) substantiated accusations against China, led to a sustained trade war. This aggressive Trump behaviour motivated the Chinese leadership to

---

[92] https://news.cgtn.com/news/2020-09-17/China-Russia-move-away-from-the-U-S-dollar-in-mutual-trade-TRCsZBLRtK/index.html

[93] https://www.thebalance.com/yuan-reserve-currency-to-global-currency-3970465

[94] https://www.globaltimes.cn/content/1186913.shtml

[95] https://www.investopedia.com/understanding-chinas-digital-yuan-5090699

[96] https://www.bbc.co.uk/news/business-54261382

accelerate efforts to strengthen their military capabilities, assert greater authority in international affairs and compete more openly in trade circles.

The post-Covid-19 era is likely to be defined by a new phase of the "Cold War", which has already begun. The contrasting approaches between the West and the East in dealing with the Coronavirus crisis has elevated China to the main challenger for the title of the world's major superpower, as all the societal and systemic weaknesses of the US and its Allies have been exposed.

In the post-Covid-19 era, everything will be much more unpredictable! We witnessed Trump and his administration vilifying China and using anti-Chinese rhetoric as a major platform in his campaign for re-election. However, with the Biden Administration assuming power, the future trajectory of this phase of the Cold War remains uncertain. The signs are not promising. From the Chinese point of view, its leaders have already made it clear that they believe the US-led West is in terminal decline, while China is on an inexorable path to a "great rejuvenation".[97]

At present, Russia appears to be a smug by-stander, observing the US versus China, Cold War 2.0. That ongoing conflict has relieved some of the pressure on Russia. It has significantly reduced the attention of the American government and media away from its activities, which has provided an opportunity to act as it wills. In any future conflict, it seems likely that Russia will side with China. Their economic ties have improved considerably, strengthening the increasing trade of Russian natural gas flowing to China,[98] while Chinese technology products are moving in the other direction.[99]

---

[97] https://en.wikipedia.org/wiki/Chinese_Dream
[98] https://www.spglobal.com/platts/en/market-insights/latest-news/natural-gas/120219-russia-starts-gas-deliveries-to-china-via-power-of-siberia
[99] https://warontherocks.com/2020/08/the-resilience-of-sino-russian-high-tech-cooperation/

## 1.5 Oil Price New Norm and Volatility

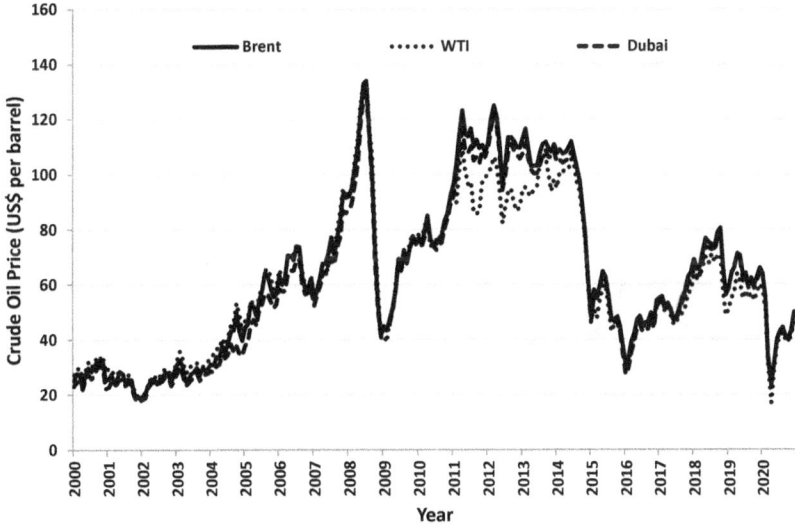

*(a) Nominal price (2008 – 2020)*

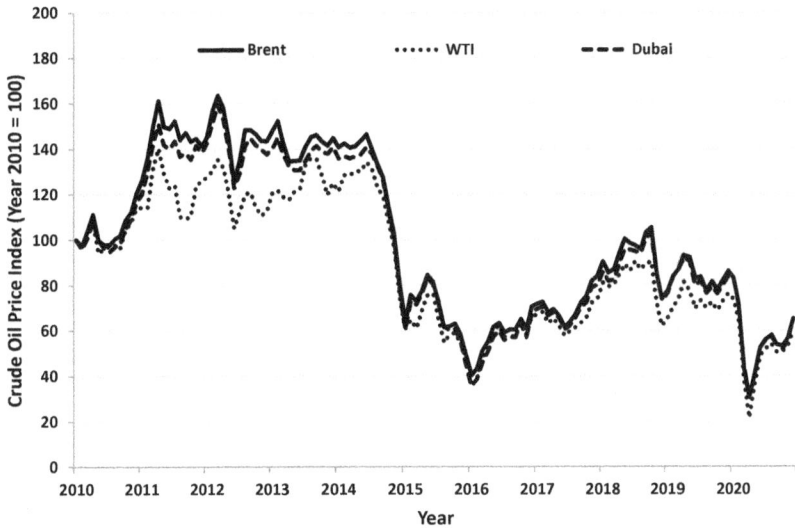

*(b) Relative price to 2010*

*Figure 1.8: Nominal crude oil price*

Source: IMF
Note 1: Prices are monthly averages.

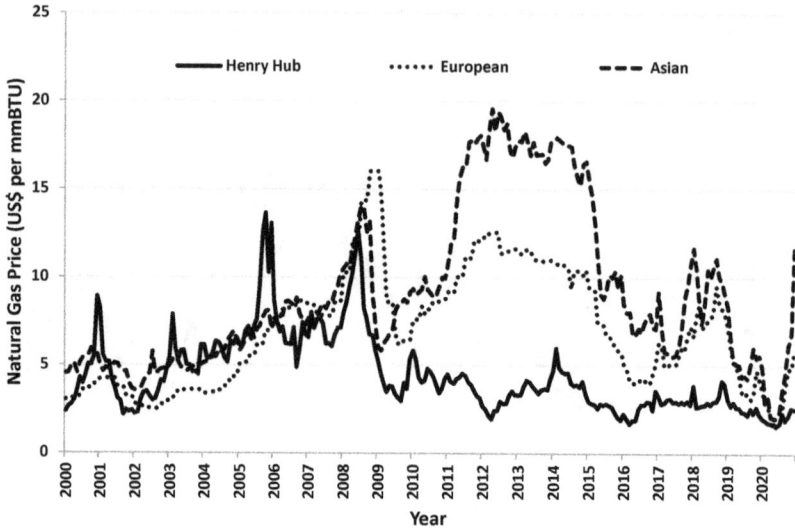

*(a) Nominal price (2008 – 2020)*

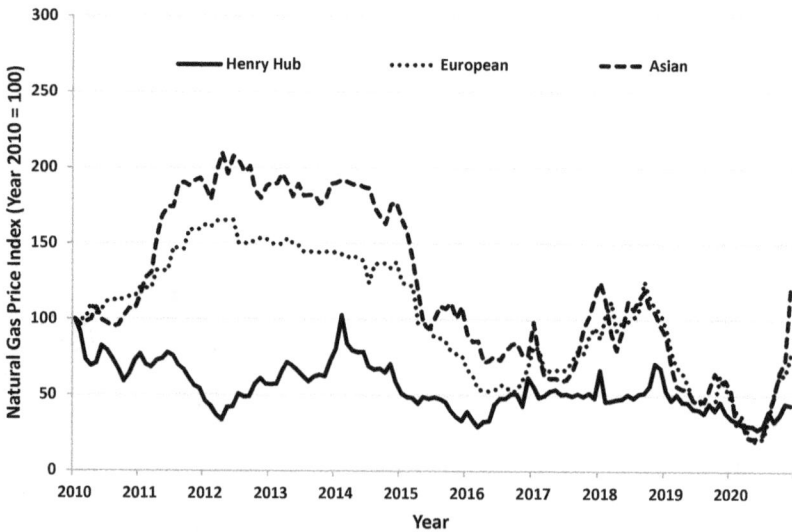

*(b) Relative price to 2010*

*Figure 1.9: Nominal natural gas price*

Source: IMF
Note 1: Prices are monthly averages.

Oil prices have experienced a rollercoaster ride over the last five years. It rose gradually from the low-price level it reached following the collapse (due to the oil supply glut of November 2014). However, it never recovered to the highs of 2014. A double "black swan"[100] of supply glut and demand destruction in April 2020, resulted in the biggest drop in oil price history. See Figure 1.8. In fact, oil price entered negative territory for a few hours for the first time in its history on April 20th 2020. Oil price is discussed in detail in Section 3.2, where future forecasts and the likelihood of lower longer-term prices are discussed.

Natural gas prices also experienced a rollercoaster, but that was distinctly different. Natural gas prices continue to be regionalised and a substantial share of the trade continues to be partially oil price indexed, with different pricing in each region. Thus, while the Henry Hub price never recovered to 2008 heights, Asian and European prices mirrored oil prices, therefore showed volatility and collapsed in November 2014 and again in 2020, although its recovery trajectory after the collapse also differed. See Figure 1.9. Natural gas price is discussed in detail in Section 3.3, where future forecasts and longer-term prices are discussed.

## 1.6    The Rise of the Information Age

Between 2010 and 2015 technology companies outperformed those in other sectors, especially the oil and gas sector. At that time this was big news. In the last five years, that gulf has grown, and the over-performance has substantially increased, dominating the market indices. This can pose a risk to the pension funds whose portfolios in other sectors are shrinking, and can have consequences on the overall global economy. Currently, the valuation of the four leading technology companies exceeds US$1 trillion, with more on a similar trajectory. In 2020, Apple's valuation exceeded the US$2 trillion mark. See Figure 1.10.[101]

---

[100] The black swan theory or theory of black swan events is a metaphor that describes an event that comes as a surprise, has a major effect, and is often inappropriately rationalized after the fact with the benefit of hindsight. (https://en.wikipedia.org/wiki/Black_swan_theory)
[101] The analysis excludes Aramco, which was privatised late 2019. The others are all technology companies.

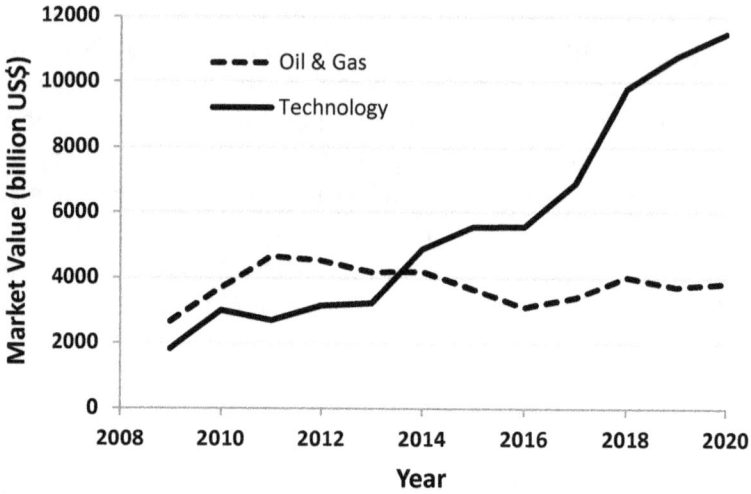

*(a) Market capitalisation (2009 – 2020)*

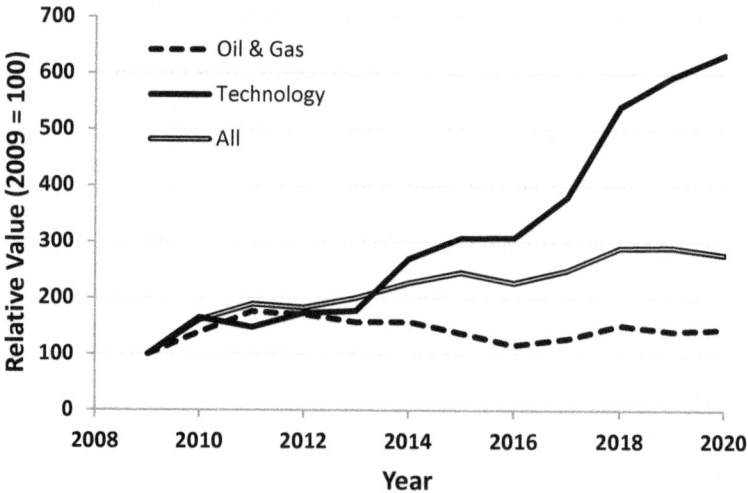

*(b) Relative to 2009*

*Figure 1.10: Comparison of technology and oil and natural gas companies*

Source: Forbes Global 2000
Note 1: Oil and gas is composed of oil and gas producers, and oil equipment and services.
Note 2: Technology is composed of software, internet retail, semiconductors & computer services.
Note 3: Includes Aramco after its IPO, which accounts for a substantial proportion of the total value of the sector, without it, the Oil & Gas sectors will nosedive to 81 in the relative value.

The technology and information industry has transformed human lives and behaviour in every area of daily life - education, health, business, science, communications, finances, engineering and other activities. This became more apparent during the pandemic, with the world living and working through the lockdowns, where the Covid-19 response was the catalyst for using technology to forge a new pathway that transitioned into a new way of living.

The view that "data is the new oil", is now widely accepted as true, with the importance and value of data fully recognised as crucial in pursuing a prosperous future. The availability of massive volumes of data and the development of "big data" analysis techniques, have transformed the way we think and utilise data to reach informed decisions or consume data. Thus, the prominence of information technology has revolutionised the world, with our ubiquitous adaptation and reliance on the internet, as well as our reliance on telecommunications. Technology has also turned human life into an Orwellian dystopia, where everything we do is monitored and recorded. The Covid-19 pandemic allowed some governments to increase the prevalence of intrusive surveillance tools and provided the excuse to impose more control over their citizens.

Just consider that by using mobile phones/fitness technology etc, all our histories are recorded by different governmental agencies and companies, most notably Google, Facebook and Huawei, who dominate the 5G technologies. Pre-Covid 19 many protested, campaigning against these intrusions, but post-Covid-19, society seems to have surrendered to the inevitability of this intrusive erosion of our privacy. When someone uses an Android handset, their privacy protection is being eroded, despite loud assurances trying to convince us otherwise. In fact, it is clear that the balance between privacy and convenience will always tip in favour of convenience, so privacy standards will inevitably decline further. Interestingly, data hacks and data leaks are now frequent, exposing both social behaviour and political dirty tricks which throws an element of additional chaos into our lives.

Benefits from new technologies, leaner processes (as a result of higher efficiencies and automation in many processes), are benefiting the companies that adopt them, giving them the competitive edge to become more dominant. New technologies, e.g. additive manufacturing (e.g. 3D printing), machine learning, artificial intelligence (AI) and the internet of things (IoT)), have reduced manufacturing production costs considerably. They have opened wide avenues for rapid future design, the development and deployment of technologies that were unthinkable even ten years ago, that will alter the way we live. These include driverless cars, connected appliances, drones, personalised medicines, etc. Some technologies have moved into full adoption stage, however the developments in AI are controversial, as some fear that it may render human beings redundant or cause a new humanitarian "winter".[102,103]

The Covid-19 pandemic illustrated the vitality of technology companies, who enabled many aspects of daily life to continue. It proved that technology companies are here to stay and diminished the prospects of splitting them in the short term. The realities of life proved that we need strong technology companies to survive and be responsive to rapid change. Thus, any concerns regarding privacy, while being publicised, are effectively being brushed aside and earlier backlashes regarding intrusion, are getting muted and dying out. The loss of privacy is the price we most probably have to pay for our current lifestyle. However, this need to be checked against ethical and moral guidelines that must be developed to ensure that tyranny through technology is prevented.

## 1.7    Fast Forward Five Years

It is depressing to read the text below from my previous book and realize that very little has changed, and that perhaps some of the

---

[102] https://www.washingtonpost.com/news/speaking-of-science/wp/2016/01/20/why-stephen-hawking-believes-the-next-100-years-may-be-humanitys-toughest-test-yet/
[103] https://www.bbc.co.uk/news/technology-51064369

world's problems have deteriorated since then.

Just reflect on the text below:

*Early in 2016, the Arab world appears to be in complete turmoil. The Arab states are further away than ever from democracy, sliding deeper into autocratic rule. The Second Cold War is at full swing. China's economy is wobbling. The oil price is in free fall. And the technology age is accelerating.*

Today I have edited the text with minimum modifications, and it reflects the current world condition. The edited paragraph below (with strikethrough and italic formatting evident to highlight the changes) describes the new reality.

*Mid 2020* ~~Early in 2016~~, the ~~Arab~~ *(not only the Arab)* world appears to be in complete turmoil. The Arab states are further away than ever from democracy, sliding deeper into autocratic rule. The Second Cold War is at full swing. *The world's* ~~China's~~ *(not China's)* economy is *collapsing and nosediving* ~~wobbling~~. The oil price is in free fall. And the technology age is accelerating.

What will happen next? A definite answer to this is impossible. Experts and analysts have already published papers. There are reports drawing up possible scenarios, forecasting events and indicators. Many of them are contradictory and certainly, as always, many will turn out to be wrong. The few who get it right will prosper perhaps. How these predictions will fare, or manifest, is anyone's guess.

Look out for the following indicators, they may give us some clues and prepare us as we move into the unknown:

- Increased entrenched polarisation in the Middle East, with Israel-UAE axis on one side and the Iran-Turkey axis on the other side.
- Saudi Arabia may struggle to navigate a safe pathway.

- A resurgence of Islamist groups with more extreme views in the Arab world, leading to a new cycle of "war on terror".
- Despair amongst the Palestinians. The loss of hope for establishing an independent state, which will lead to either major strife or the acceptance to either Apartheid conditions or a form of one-state solution.
- Another cycle of low oil price, with prices stuck in a narrow band. This could cause major unrest in the Middle East region, as governments will struggle to cope with the subsequent economic hardships.
- A more confident China, following its success dealing with Covid-19 pandemic, leading to intensified conflicts with its neighbours (in South China Sea and India) and on the global stage, with the US, destabilising world politics.
- A wave of nationalism and deglobalisation following the economic downturn caused by the incompetence of the world dealing with Covid-19 crisis. This will lead to the re-emergence of neo-fascism and populism in many countries. (The Trump administration was just an audition to this new world theatre)!
- Major civil unrest in the US, leading to the emergence of splintered secessionist movements. Along with the waning of US influence globally.
- The future of the EU following Brexit. Will other member states choose to exit – Italy, Sweden or the Netherlands? Failure to progress will threaten those states with Russian interference and/or the return of fascist governments there.
- Policies fighting climate change turning from "aspirational pledges" to bold urgent action, including the wide adoption of electric cars/public transport. Consider the knock-on effect of that on the Middle East oil and natural gas trade.

# Chapter 2
# SIGNPOSTS – WHERE DO WE STAND?

In my previous book "Fossil Fuels in the Arab world: Seasons Reversed",[1] I listed several events to watch out for which I predicted would shape the Arab world's future and which would have global implications. At that time the events I listed were classified into two groups that have direct, or indirect, consequences on the Arab world.

The top three signposts from an Arabic perspective were:
- Saudi Arabia instability risks, discord within the royal family and possible conflict with Iran. This could cause an oil price shock, which would lead to a world economic recession.
- A growth in ISIS activities, inside the Middle East and extending its reach around the world. This would increase terrorist attacks worldwide, with far reaching implications on travel, work and immigration.
- The oil price scenarios establishing when the new oil price floor will be reached. Is this floor below US$30, around US$50 or higher?

The other three signposts were outside the Middle East, but their ramifications would be felt strongly in the region:
- China's economic performance, detecting possible wobble in the financial markets, as well as its actions in South China Sea. This would have major economic consequences globally, as well as increasing fears about military conflicts in the region.

---

[1] Basel Nashat Asmar, Fossil Fuels in the Arab World: Facts and Fiction – Global and Arab Insights of Oil, Natural Gas & Coal, 2050 Consulting, London, UK, 2010

- Russia's foreign policy, especially in Ukraine and the Middle East. How would the current Cold War be seen around the Globe during the next 8 years?
- The future of the EU and its possible fragmentation, considering the policy development in response to the immigrants' crisis and the Brexit aftermath: risks, impacts and implications.

Analysing the timeline of the above events gives us clear insights to where we are heading. However, we need to differentiate between events that have happened and had an outcome, to the ones that are still unfolding.

Below I look at each of the six signposts in more detail.

## 2.1   Saudi Arabia Instability Risks

In the last five years Saudi Arabia has undergone significant transformation. The magnitude and pace of change took everyone by surprise, altering many traditional rules and cornerstone policies.

In the following paragraphs there is no judgement on the specifics of the actual policies and their merits. The text below is simply examining what has happened and the consequences. There is a risk assessment, and we ask how these changes affect stability in the country and the region.

Recent events have affected the country's political, financial and social norms, where the power of influential individuals was curtailed, and new power houses emerged. This shift in power has disturbed the foundations of the Saudi state. It has spread uncertainty, anxiety and fears at home. Meanwhile it has confused allies and foes externally, leaving them scrambling to respond to these unprecedented changes. Combing through the reams of analysis written prior to 2017, it is clear that experts were completely surprised by the rapid pace of development and the fundamental changes.

The main trigger for this change was the rise of Mohammed bin Salman, widely known as MBS. In January 2015, he was appointed as Defence Minister and Deputy Crown Prince. He then, in a rapid ascendance, tightened his grip on power. He assumed the de-facto position of absolute ruler of Saudi Arabia in June 2017, when he was officially elevated to the position of crown prince.

Initially his stratospheric rise was met with great optimism and enthusiasm, especially amongst young people. Domestically he initiated economic reforms, restricted the powers of religious police, liberalised cultural laws, allowing cinemas to open for example and removed the ban on women driving. He portrayed himself as a reformer, with local, regional and international media covering stories of this radical charismatic figure. He appeared on major magazine covers and many world leaders wanted to be seen associating with him.

He announced the "Saudi Vision 2030", which was hailed as a strategic framework to reduce Saudi Arabia's dependence on oil by diversifying its economy to expand other sectors. It envisaged the development of public service sectors such as health, education, infrastructure, recreation, and tourism. At the height of his popularity, he was seen as a herald of positive progressive change in his country.

However, success in implementing the reforms has been limited. Internal opposition obstructed the smooth implementation of the new ways. The innovations ambitiously tried to change the minds of the population, taking on the huge challenge of reversing traditional thoughts and ideas, but in fact, they left many conservatives dismayed. Thus, only partial reforms have managed to take place. Corruption, bribery, red tape and time-consuming bureaucracy continue to stifle the Saudi system. This has sabotaged the meaningful reforms needed to restructure the economy and wean it off its dependency on oil. The public has witnessed only selective eradication of corruption. There was the high-profile purge of opponents - the infamous Ritz-Carlton incident, where a number of elite figures - prominent Saudi Arabian princes, government ministers, and businesspeople, were arrested in November 2017, accused of corruption. Following a lengthy investigation, over US$100 billion were retrieved. Despite a result that appeared to be a vindication of MBS efforts, the incident destroyed decades of internal harmony in the Kingdom. It also damaged the confidence of both private and foreign investors fearful of pumping money into an unstable market, severely reducing the inflow of foreign investment.

In the tourism industry, changes were also made to make Saudi Arabia as attractive a holiday destination as other Arab countries. There were huge marketing campaigns to promote the attractions of the country. Additional reforms were made, such as the rapid openings of cinemas, concert halls and other entertainment venues – all of which would make the country more inviting for international travellers. In addition, there was some success with the introduction of tourist visas, making it easier to travel into the country from abroad.

However, the country's image changed dramatically after miscalculations and incidents. The assassination inside the Saudi Consulate in Istanbul in October 2018 of Jamal Khashoggi, a Saudi dissident and journalist for The Washington Post, dealt a massive public relations blow. It highlighted the increased human rights abuses of the new era, where opposition and dissent are not tolerated. There was widespread international condemnation of the brutal murder. As the outcome of investigation into the incident was awaited, there were concerns that the Saudi sovereign wealth fund, known as the Public Investment Fund, chaired by MBS, was now vulnerable to sanctions internationally. Internal critics feared that MBS' recklessness and willingness to both risk Saudi assets and put the country in international crosshairs for his personal agenda placed the Saudi economy in danger. The international private sector shared concerns for the implications of the atrocity and its accompanying publicity, on their activities – how would they proceed if Saudi became an international pariah?[2]

This catastrophic political impact of this heinous act was in full view at the G20 Summit in Argentina in December 2018. The Crown Prince was shunned like a pariah. World leaders kept away from him and tried to avoid being photographed with him.

The slow pace of reforms, the brutal murder of Khashoggi and the purge of opponents at the Ritz Carlton, all cost MBS enormously. He lost his reputation as a reformer. The legitimacy of the whole royal family was threatened when he revealed the extent of the open family infighting. Their legitimacy was further threatened when he initiated a split with the religious Wahhabi establishment, clipping of the power of the Committee for the Promotion of Virtue and the Prevention of Vice (locally known as *mutawa* or *mutaween*). The repercussions of this conflict shook the very foundations of the stability of the kingdom.

Internationally, the Saudi government pursued more forceful policies, with increasingly aggressive involvement in the wars in Syria, Libya and Yemen. The latter is now seen as Saudi Arabia's Afghanistan. The Saudi government appears to be stuck, even though they realise that their action in Yemen is unwinnable and it is costing them a fortune, they need to save face. They believe that a retreat will embolden the Houthis, who are attacking deep inside the kingdom, and leave their border exposed to their archenemy, Iran. See Section 1.3.1.

Moreover, their aggressive foreign policy fuelled widespread regional instability with - a chaotic blockade of Qatar in tatters after 3.5 unfruitful years; a failed forced attempt to remove the Lebanese prime minister; seeding a possible future conflict with Egypt by grabbing the islands of Tiran and Sanafir; and open hostility with Iran, to the degree of encouraging President Trump, (who was admired greatly by the Saudi government), to start a military conflict. The support for other Trump policies is evident

---

[2] https://www.cnbc.com/2021/02/26/saudi-fund-vulnerable-after-mbs-actions-in-khashoggi-killing-ex-obama-official-says.html

in their attempts at semi-normalisation of relations with Israel and the promises of massive weapons deals with the US. The farcical reception for Trump and his daughter Ivanka during his first official visit to the kingdom, when Trump subsequently boasted (internally to his supporters) that he would extort money from the Saudis in return of providing protection. Following Trump's loss and the imminent change in the American administration, the Saudi regional policies will have to be reassessed to conform with new American priorities.

In 2016 MBS announced the intention to privatise the Saudi national oil company, Aramco. Initially, the idea of Aramco's privatisation appeared to be a shrewd move to improve its governance and eliminate corruption by forcing transparent disclosure of activities and expenses. It would have the additional advantage of raising a substantial amount of funds. Alas! The implementation went badly.

The saga of attempting to privatise Aramco demonstrated to the world that transparency in the Saudi market remains a mirage. The unpredictable changing of rules and the reluctance of the leadership to effectively manage the change, led to extended delays that dragged for several years, eroding any enthusiasm there may have been for the project. MBS insistence on a valuation of US$2 trillion, which most experts felt was too high, complicated the matter. Aramco's initial public offering (IPO) was marketed as a great opportunity to attract foreign investment. The aim was to list 5% of the company, raise US$100 billion, and list the company on foreign stock exchanges. However, with the affair faltering as a result of legal threats (e.g. the Justice Against Sponsors of Terrorism Act (JASTA) of 2016,[3] and the proposed No Oil Producing and Exporting Cartels Act – known as the NOPEC bill[4]), meant that the shares on offer were reduced to 1.5%, and the Saudis were forced to downsize their aim and only offer a local listing. They achieved one public relations objective

---

[3] https://www.nortonrosefulbright.com/en-de/knowledge/publications/d1a384e4/laypersons-guide---justice-against-sponsors-of-terrorism-act
[4] https://en.wikipedia.org/wiki/No_Oil_Producing_and_Exporting_Cartels_Act

though, with the IPO becoming the largest ever historical IPO by a whisker,[5] raising US$25.6 billion and increasing to a record US$29.4 billion, after selling more shares. But all the shares were offered to local and selected GCC citizens, and many share sales were forced on the Ritz-Carlton detainees!

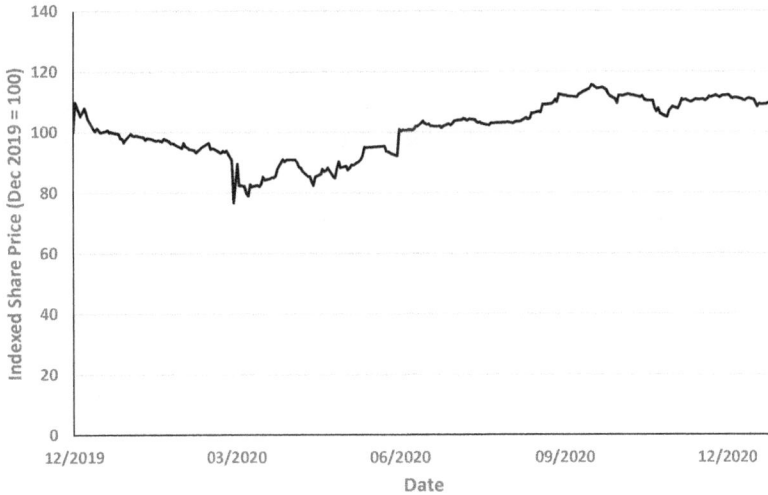

*Figure 2.1: Aramco share price relative to IPO data*
Source: Google Finance

The IPO valued the company at US$1.6 trillion. In the days following the IPO, to fulfil MBS initial objective and boost his ego, market manipulation raised its valuation to US$2 trillion. But the inflated valuation was only temporary, as the shares dropping down again. See Figure 2.1. At the time of writing (December 2020) Aramco was valued at US$1.87 trillion,[6] making it the second highest value listed company in the world, behind Apple. However, due to oil price collapse, its revenue declined significantly in March, although it trended upwards since then, as a result of improving oil price. Aramco had to

---

[5] https://www.cnbc.com/2020/01/12/saudi-aramco-raises-ipo-to-record-29point4-billion-through-greenshoe-option.html#:~:text=Saudi%20Aramco%20raises%20IPO%20to%20record%20%2429.4%20billion%20through%20greenshoe%20option,-Published%20Sun%2C%20Jan&text=Aramco%20initially%20raised%20a%20then,the%20over%2Dallotment%20of%20shares.
[6] https://markets.businessinsider.com/stocks/aramco-stock

borrow US$10 billion after the oil price collapse of 2020 to maintain its promised dividend.[7]

Both internal reform and aggressive foreign policy added pressure to the balance sheet of the kingdom. It was forced to try to bolster its revenues by triggering an oil price war, in the hope of maximising its oil revenue. However, both attempts in 2014 and 2020 backfired spectacularly, creating economic hardship. Following the first botched oil price war, Saudi Arabia reluctantly reached a truce with Russia, in the framework of OPEC+, leading to a relative success controlling oil price. Then, in 2020, the second oil price war failed miserably when the demand collapse due to the Covid-19 pandemic was initially underestimated. Thus, although the supply glut was the responsibility of Saudi Arabia's actions, the resulting demand collapse was not. So, Saudi Arabia is only partially responsible. At the time of writing (December 2020), with Trump threatening Saudi Arabia directly for triggering the oil price war,[8] a comprehensive agreement involving Russia, OPEC and other producers, appears to be a success for Saudi diplomacy, bolstering oil prices.

It is important to realise that in a price war, Russia has the advantage over Saudi Arabia, since it can devalue its currency, while Saudi Arabia and other Gulf countries cannot, as their currencies are tied to the US$.

This decline in revenue has caused a second, four-year long, period of austerity.[9] This has forced the Saudi government to increase the value added tax from 5% to 15%. It has had to remove the cost-of-living allowance and curtail its public spending. This, in turn, will force a reduction in future development plans, especially Vision 2030 and will make its delay inevitable.

---

[7] https://www.hartenergy.com/news/saudi-aramco-closes-10-billion-loan-187545

[8] https://uk.reuters.com/article/us-global-oil-trump-saudi-specialreport/special-report-trump-told-saudi-cut-oil-supply-or-lose-u-s-military-support-sources-idUKKBN22C1V4

[9] https://www.bloomberg.com/news/articles/2020-05-11/saudi-arabia-plans-26-6-billion-austerity-cuts-triples-vat-ka1uss4c

At the moment the financial crisis, triggered by the second oil price war and subsequent collapse, is making Vision 2030 appear totally unrealistic. In addition, the systematic violation of human rights – discouraging foreign investment, and the tensions and infighting within the Royal family (actually tribe) shattering its illusion of unity, are complicating the prospects as well. With a massive budget deficit (as a result of the low oil price) and the collapse of tourism income due to first the suspension, then the restrictions of umra (small pilgrimage) and severe curtailment of hajj (pilgrimage), the Saudi government will be forced to borrow heavily to fulfil its fiscal obligations. It is therefore likely that other visionary projects, such as Neom[10], will be postponed or halted in the shorter term, despite repeated announcements that the project is still forging ahead.[11]

The stretched fiscal austerity will threaten social cohesion in the Kingdom, increase poverty, tensions, affecting its prospects and stability. In the past, the Kingdom used its sovereign fund as a last resource to get out of trouble. But with the fund declining significantly in the last two years, partially by transferring to PIF, this time it may not be possible. See Figure 2.2.

In fairness, the economic reform in Saudi Arabia, initiated by MBS has started, but it is still limited, and further bold, true economic reform is needed in order for the country to move forward. Economic reform alone is not enough to move the country away from oil dependence. This goal will only be achieved if it is accompanied by legal reform and a framework for gradual political reform.

Time will tell if MBS is a genius who saw the writing is on the wall, realising that, not only is  the oil decline terminal, but that the world is changing rapidly especially in the Middle East, and who seized the moment to change his society, or a fantasist whose vision was built on sand.

---

[10] Neom is a is a planned futuristic cross-border city in the Tabuk Province of northwestern Saudi Arabia. It is planned to incorporate smart city technologies and function as a tourist destination.
[11] https://www.middleeasteye.net/news/saudi-arabia-the-line-neom-megacity

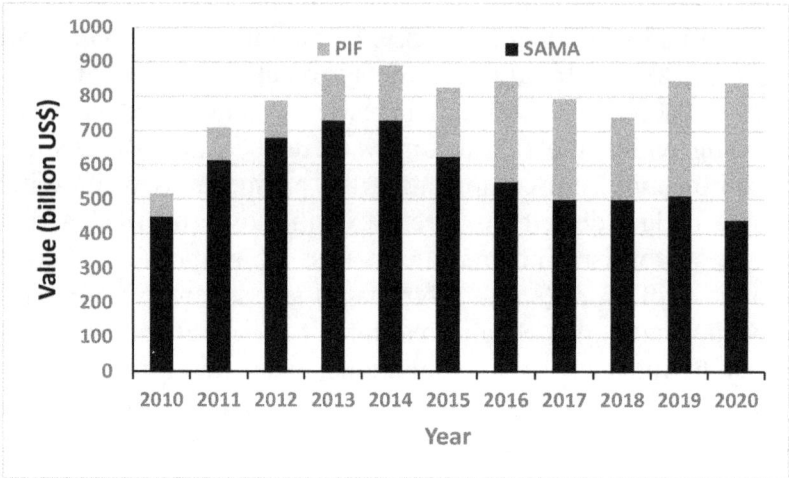

*Figure 2.2: Saudi SAMA and PIF (2014 – 2020)*

Source: The Economist; Sovereign Wealth Fund Institute
Note 1: SAMA is the Saudi Central Bank; PIF is the Saudi Arabian Public Investment Fund

Ironically, as a cruel form of karma (or schadenfreude)[12], when the unthinkable decisions in Saudi Arabia liberalising the society started taking hold, the Covid-19 pandemic arrived, bringing a sudden halt to many changes.

## 2.2 ISIS Collapse

The rise of ISIS began in 2014 as a result of the weakening of the Syrian army, when a power vacuum was successfully exploited by the group. This particular Islamic extremist rebel group, declared an Islamic Caliphate, controlling large swaths of Syrian land - approximately 50%. At that time, other Islamic rebel groups as far as Nigeria and the Caucasus, pledged allegiance to the organisation. In addition, terrorists, allied to the group, attacked Western targets in Europe and North America, which highlighted the serious threat extremist groups pose to the West and its allies.

---

[12] Arabic شماتة

By November 2015, in addition to its Syrian territorial control, ISIS controlled roughly 30% of Iraq's land in the west and northwest. At its height it controlled major cities including Mosul, Ramadi, Tikrit and Raqqa, and had over 10 million people under its control.[13]

This Iraqi territory included several oil fields, which allowed ISIS to generate income from the black-market oil trade and discouraged international oil companies from pursuing further opportunities in Iraq.

The growth of Daesh, aka Islamic State (IS), ISIS or ISIL[14] as a pseudo-political entity, after peaking in 2015, started reversing and came to a halt.

In 2017, following a sustained military campaign by Arab and Western governments, as well as other uncoordinated attacks by Russian and Iranian forces[15] ISIS lost its territory, and was reduced to few pockets, mainly in isolated areas in the Syrian desert. Its leader disappeared from the public eye and then was killed in 2019. Although the head seems to have been cut off this ideological snake, due to the continuing power vacuum and the weakness of central governments in both Syria and Iraq, a resurgence is always possible, albeit the probability is remote.

In addition to the losses in Syria and Iraq, ISIS also lost their territories in Libya and Sinai, but remains active in many countries, mostly operating underground. At the moment, ISIS is defeated but it is certainly not eradicated and the threat of radicalised foreign fighters from within its ranks, returning to their countries of origin and destabilising them, remains a real possibility.

The Covid-19 crisis appears to have created a new opening, with power vacuums again spreading in many areas, allowing ISIS to

---

[13] https://www.bbc.co.uk/news/world-middle-east-27838034
[14] ISIS stands for the Islamic State of Iraq and Syria, while ISIL for Islamic State of Iraq and the Levant.
[15] https://en.wikipedia.org/wiki/International_military_intervention_against_ISIL

regain some strength. However, for the time being, while it can stage sporadic attacks, it remains relatively weak and controls little territory.

Note however that some in the Middle East view the whole ISIS story with scepticism and consider it as part of "false flag" conspiracy campaigns waged by foreign powers, e.g. USA. They believe that ISIS was created by the West and that they killed the leader Baghdadi, who they accuse of being an American or Israeli agent.[16] In the last few years, photoshopped photographs went viral on the WhatsApp platform, claiming to confirm these conspiracy theories. However, these were, in fact, examples of fake news being spread in the region.

## 2.3 Oil Price Volatility: Recovery then Collapse

I discussed briefly in Section 1.5 the rollercoaster price volatility in the oil markets. The 2020 "floor" reached was never envisaged before and the ceiling for the future remains up in the air following the double whammy of supply glut and demand destruction.

While prices recovered in May 2020, following a historic collapse of the oil price in April 2020, prices remain volatile. The success of the accord curtailing production between OPEC and non-OPEC oil producers, via the OPEC+ framework, overachieved in the short term, but the prospects of long-term success controlling prices remain uncertain.

One important takeaway from the latest oil price collapse is that it has confirmed that oil price will have a ceiling. It will never reach old high levels and there will be a low ceiling in the long term, but with sporadic spikes both peaks or troughs along the way.

An interesting piece of trivia is that the negative oil price occurred on the tenth anniversary of Deepwater Horizon disaster

---

[16] https://www.gospanews.net/en/2019/08/06/al-baghdadi-issi-caliph-and-cia-agent-hidden-by-us/

in the Gulf of Mexico.[17] It is probably a coincidence, but it is unlikely coincidences like this that are the foundations of conspiracy theories so who knows what may be imagined pointing to the contrary in the future?

For full discussion on the causes and implications, see Section 3.2.

## 2.4   China's Economic Decline is Manageable

In Section 1.4.2 I discussed the events involving China in the last five years, the current situation there and the future implications in detail. Here I just re-iterate few points that summarises the main takeaways.

Firstly, China is manipulating the media at home and abroad, spreading anti-USA propaganda across all formats. It is using a distorted version of history to demonise its rivals and justify its current ambitions. Trump's ostensibly "Republican" administration undermining of the US democratic principles, was cleverly employed by China to promote its own authoritarian philosophy. However, by promoting nationalism, China conceals the dysfunctional malaise of its governance, along with the other structural weaknesses it experiences, especially economic, from its population. China may follow the Russian strategy, using the welfare and the safety of Chinese minorities in neighbouring territories, as a pretext to interfere in the affairs of other countries. It continues hostilities with India, fermenting a conflict based on ancient historic territorial claims.

Secondly, the decoupling of China from the US is both irrevocable and accelerating. This situation, in turn, poses challenges to the weakening power of Europe. Which side to

---

[17] An explosion on the drilling rig Deepwater Horizon at the Macondo Prospect (Mississippi Canyon Block 252) occurred on April 20, 2010, killing 11 workers and resulting in a massive oil spill and environmental disaster. The Deepwater Horizon sank on April 22, 2010, in water approximately 1500 m deep. Following the rig explosion and subsea blowout, it took BP three months to drill a relief well. The well was successfully sealed off from flow into the sea on August 4, 2010 by a "static kill" (injection of heavy fluids and cement into the wellhead at the mudline; https://en.wikipedia.org/wiki/Macondo_Prospect

align itself with? How it will respond to the new "Silk Road" policy – i.e. Communist China's softer economic and diplomatic policies? Although the decoupling is hurting China, ultimately it will benefit in the longer term, by helping the economy to maintain respectable growth rates, concentrating on internal markets and becoming self-reliant. It appears that the West is exaggerating the extent of Chinese economic woes.

Thirdly, China has fared better than other countries affected by the Covid-19 pandemic. It has technically avoided recession, while most Western countries' economies have suffered. This will tip the scales of power further in the direction of Asia Pacific and hand the advantage to China. Its response to the pandemic allowed China to become more assertive in pursuing its international affairs, especially in Africa and Latin America. It provided China with an opportunity to replace its "loans provision soft colonialism" policy with an alternative "vaccine colonialism" policy, by offering vaccines to Africa, while highlighting the selfish Western nations practice of stockpiling vaccines and withholding production independence from poorer nations. In time, it is possible that we may see an overt Chinese military presence, including the building of military bases, in some African and Asian countries.

The Covid-19 pandemic and its global aftermath have given China more confidence. It will determine the position of China on the world stage for years to come. Thus, in the next few years we can predict some of the steps China will take to empower itself further:

1. Implementing actions to erode democracy in Hong Kong (and Macao) and imposing more direct Chinese involvement.
2. Accelerating "Sinofication" in its outer provinces, especially Xinjiang.
3. Employing provocative policies, exerting more pressure and harassing Taiwan.
4. Asserting sovereignty over South China Sea by

intimidating other countries with territorial claim disputes. Also pursuing more aggressive polices to change facts on the ground and administratively creating new districts in the territory, e.g., construction of new islands.

5. Strengthening its military power, especially the air force and the navy, by acquiring new aircraft carriers.

6. Orchestrating skirmishes with India and intensifying discussions about China's historical territorial claims with neighbouring countries and its "proper" place in the world.

7. Strengthening its strategic alliance and cooperation with Russia.

8. Advancing its position as a technology leader, especially AI and telecommunications. Also, exploiting the backlash against American policy to justify even more censorship and restrictions on the internet, thus leading to fragmentation of protest at home.

## 2.5 The awakening of the Russian Bear

In Section 1.4.1 I discussed the events involving Russia in the last five years. Briefly, I assert that the Russian bear has been awakened and is reasserting its previous powerful position on the world stage. In Putin, we can see a confident leader, willing to do as he wishes, despite international protests. We can learn some important lessons:

1. Russia's expansionist policies have proved successful, with the Crimea becoming an integral part of Russia, despite the cries and objections of the West. Thus, it is likely that more expansionist attempts will be made.

2. The Western sanctions, which initially hurt, ultimately proved fruitless and ineffective. Despite the sanctions, the Russian economy forged ahead with substantial oil revenues, prior to the 2020 oil price collapse.

3. Putin is here to stay. He can stay in power till 2036 and, unless fate interferes before that,[18] the world has to accept

---

[18] https://www.cnbc.com/2020/07/02/russia-vote-victory-for-putin-who-could-now-stay-in-power-

this fact.

4. Russian foreign policy is becoming more aggressive. Russian troops are stationed abroad, not only in the former Soviet Union, but also Syria and Libya. It may establish itself further afield.

5. Russia has mastered cyber and electronic warfare and is using these techniques effectively all over the world. High profile hacks in the US are frequent. This cyber-conflict will only intensify.

6. Russia is spreading anti-US propaganda and enlisting China's help in pursuing this strategy.

7. Russia is taking a leaf of China's book and is pursuing its own separate censored internet. It is forcing Western companies to accept its conditions in order for them to be allowed to work in the country.

## 2.6   The Ineptness of the EU

The last five years have rocked the foundations of the European Union to its core and the Block of 27 countries has faced numerous crises: political, financial and moral. The EU is perceived to have been generally slow in responding to these various issues, with infighting between its member states rendering its responses weak and ineffective. This has left many observers questioning the future of the Block. It has proved to be inept with some of its decisions, but, if history tells us anything, it is that the EU is a survivor, e.g. the Greek financial meltdown and the migrant crisis in 2015. Currently, the Block is facing three main issues, which will define its future.

First is the growing nationalism in almost all its countries. There has been divergence from democratic rule in some of its member states, with Poland and Hungary leading a group of countries whose populist leaders are challenging the accepted European rules. In addition, there is growing social unrest in many member states. In Italy, Spain and France, new groups have formed in response to social conditions, i.e. the Yellow vests movement,

until-2036.html

with growing separatism in Spain such as the Catalan independence movement. Similar developments are occurring in the UK especially after Brexit with growing separatism movements such as e well established Scottish movements and the slowly emerging and Welsh movement.

The second larger crisis facing the EU, is its handling of the Covid-19 pandemic with its subsequent economic fallout. Here the Block demonstrated its dis-union. Many governments acted in very narrow, selfish, nationalistic ways, pursuing policies to protect their borders and in some cases, suspending the Schengen free-movement area, a cornerstone of free movement European policy. The Covid-19 pandemic exposed a vulnerability of the Block and demonstrated that, in a major crisis, the mission of having a coherent policy or response, is fragile. The pandemic has delayed further expansion of the EU in the short or medium term. At the time of writing, with a new variant wave of Covid-19 spreading in Europe, a return to normal is a distant prospect. It will take brave leadership and a paradigm shift towards true international cooperation, to restore belief in an ideal European unity. A further obstacle in the near future, with respected key players like Angela Merkel, stepping down – is the lag period until a new dynamic political leader (not Boris Johnson – just kidding) emerges to lead the way.

The EU is not out of the woods yet dealing with Covid-19. Its rollout of the vaccination program was uneven at best. There has already been a devastating wave of the Delta variant. What will happen if the world suddenly has a new variant travelling fast throughout Africa or South America? If there is a new refugee crisis, will happen if Turkey opens its borders, once again allowing the migrants to enter Europe? Effective future planning will require practical, realistic policies, with broad global and international cooperation to manage such challenging humanitarian aspects of the pandemic in the future.[19] Indeed, the

---

[19] There are university courses in Disaster Management and I am sure there are government agencies whose raison d'etre is to plan for 'what if' scenarios – biological warfare/natural disasters/man-made disasters. Perhaps the solution for the EU is to open that up, like InnoCentive (an open innovation and crowdsourcing company) and ask the people who might actually know.

same will apply for climate related catastrophes also.

Third, the belligerent politics currently in the ascendancy in the UK continues to threaten the EU. The aftermath of Brexit will be crucial in shaping the EU's future. Will it hold together, or will it fragment? A successful UK, outside the EU, might encourage other member states to consider leaving. The EU hardened its positions to teach the UK a lesson and present its hardships as a lesson to other wavering members of how not to do things and the bad consequences of leaving, even with a bad deal. Hardships suffered by the UK will be for all to see.

In 2020 public opinion in the UK warmed towards re-joining the EU even before the end of the transition period. Another membership application is indeed possible, but certainly not in the short-term. Expect it to become a policy of one of the major parties in the future, when the public experience the drawbacks of the bad agreement the UK government delivered. This saga will continue for years to come.

With the UK finally out of the EU, expect more infighting and tensions to surface within the Block. France can no longer blame the UK of being the chief saboteur anymore, so we may see conflicts between France and Germany, or between the big two with smaller nations. This may provide an opportunity for nationalist or far-right movements to exploit those rifts, undermine the Block and demand withdrawal from the EU. Maybe soon we will hear (the pitter patter of Italian, Dutch or Swedish feet heading for the exit?) of Italexit, Nexit or Swexit from Eurosceptics.

## 2.7   Did We Hit the Yardsticks?

Having revisited the six signposts, I can answer that most of the yardsticks were reached. However, while this implies foresight or predictability around many of the topics, actually most of them

---

Although we have seen how someone like Trump can wreck even the best laid plans for pandemic responses.

were predictable if people put their minds to it.

Nevertheless, all of these events diminish in importance, when measured against the impact of the unexpected Covid-19 virus and the pandemic's effects on the world.

The pandemic has forced people and governments to pause and reflect, facing crucial questions: "Where do we stand? What positions are we in? And how do we find a way forward and navigate a path ahead? Do we need to work the way we have been working? Do the developed countries have an obligation to vaccinate poor countries? Are they obliged to repair the environmental damage inflicted on poorer nations?"

The answers to these questions are not straight forward and are up for debate. There are many conflicting views, all of which require the evaluation of possibilities, opportunities and risks.

When I focus on the core subject of this book and narrow these questions down to that, two main themes emerge concerning global future and the Arab world's place in a new global system. I present these two questions below and devote the next few chapters trying to answer them.

Question One – As energy transition accelerates, will the Arab petrostates be the last to give up their oil-based energy industry?

Question Two - Has the Arab world "missed the boat" and is it too late for Arabs to capitalise on the new energy revolution, using their coastlines, and sun-drenched landmass to provide exportable cleaner energy? Otherwise, they may end up with worthless oil and natural gas left underground, where they will be unable to utilise or monetize.

One of the most significant issues and unanswered questions is why is the problem of overpopulation not being addressed? This is despite it being clearly the fundamental problem at the root of running out of fossil fuels, and high greenhouse gas emissions.

## Chapter 3
# GLOBAL ENERGY SUPPLY, DEMAND AND THE TRANSITION PUZZLE

Energy supply and market dynamics gravitate around forecasting. Many commercial and political decisions are taken, based on numerous forecasts of energy supply and demand, that affect the way the world economy functions. However, forecasting is an art not a science. This is a fact, despite assertions by institutions and organisations that regularly publish forecasts to the contrary. It is true that forecasters developed models that can often predict the future based on a set of variables, but these forecasts are often uncertain. While many planners pay attention to all these forecasts, those making the forecasts have a habit of revising their model assumptions often, so that their model results also change.

Let's compare energy forecasts from IEA or OPEC for the last five years. When we look at Figure 3.14, we can see how different these forecasts are, even without taking into consideration the impact of Covid-19 pandemic. COVID-19 forced planners to issue new forecasts, with very different outcomes and to revise them more frequently.

In this chapter I discuss the market fundamentals of oil and natural gas. I look at the new era of rapid energy transition and its fossil fuels demand, which will continue to dominate the energy landscape for many years to come. I will also identify the main beneficiaries of the rapidly transforming oil and natural gas markets along with their positioning in a new energy map.

A fundamental issue to explore in order to appreciate the markets transformation, is that of fossil fuel prices and their interactions with other market factors.

In order to understand the markets dynamics and position the Arab world in the new energy order, we need to focus on the essential data for fossil fuel prices and reserves. This investigation is carried out firstly on a global scale (See Section 3.4) and is then narrowed to focus on the Arab world specifics to position it within a global perspective. See Sections 4.1 and 4.2.

In this context, geology is not the only issue. There could also an equally important, or even bigger issue, affecting the Arab world, namely how the worlds producers are scared away from investing in oil and natural gas due to Environmental, Social, and Governance (ESG) concerns. The question becomes whether the Middle East can secure financing to fill the gap. Or – what carbon price will be applied eventually to oil and natural gas production and how will that impact demand and investment?

Returning to my previous book, a fundamental question I addressed reappears, as we are living, once again, in a low oil price environment. That question is, "Who benefits from fossil fuel price oscillation, volatility and instability?"

Answering this question is vital, since fossil fuel prices are essential in determining the volume of reserves; where a higher price can significantly increase the reserve volume or endowment, while low price can drastically lower the volume or endowment.

It is necessary to understand the factors underlying the two pillars of fossil fuels market fundamentals - price and volume, which are quantified by reserves and resources of fossil fuels. We need to understand the intertwining linkages between them; the linkages with carbon policy and pricing; shareholder or customer sentiment, which also impacts on oil and natural gas consumption. In the three sections below, these factors are discussed briefly.

## 3.1 Fundamental Fossil Fuels Terminology: Understanding the Terms "Resources" and "Reserves"

Here, I need to revisit an important subject from my previous two books.[1,2] Although I may be repeating earlier writing, I believe is essential to provide a thorough understanding of this area so that the subsequent events and possible consequences, are as clear as possible to my readers.

The media has a habit of incorrect and loose usage of terminology and definitions. This deliberate, ignorant, intentional or unintentional use of a mix of technical engineering and geological terms, often leads to misinformation and inaccurate conclusions.

While the text in this section is not a dictionary of all these terms, it will provide clarity about the two main terms often mixed up when discussing or reporting oil and gas issues, i.e. "reserves" and "resources".

The definitions of fossil fuel resources and reserves are summarised in this section. If you do not already have a copy in your library, for a more comprehensive discussion, please refer to my previous book.[3]

In the most simplistic terms, *fossil fuel resources* are defined as the fossil fuels initially-in-place, i.e. the quantity of fossil fuels, in varied sized deposits and reservoirs, that exist originally on, or under the earth's crust, in naturally occurring accumulations. These resources "initially in place" are referred to by numerous names including *resource base, hydrocarbon endowment,* or *original-in-place.*

---

[1] Basel Nashat Asmar, Fossil Fuels in the Arab World: Facts and Fiction – Global and Arab Insights of Oil, Natural Gas & Coal, 2050 Consulting, London, UK, 2010
[2] Basel Nashat Asmar, Fossil Fuels in the Arab World: Seasons Reversed – Oil and Politics Interplay in the Arab World, 2050 Consulting, London, UK, 2017
[3] Basel Nashat Asmar, Fossil Fuels in the Arab World: Facts and Fiction – Global and Arab Insights of Oil, Natural Gas & Coal, 2050 Consulting, London, UK, 2010

Two main criteria categorise fossil fuels resources: discovery and recovery. Accordingly, in terms of discovery, fossil fuels are either discovered or undiscovered. In terms of recovery they are either recoverable or unrecoverable.[4] Recoverability is assessed at present day technology and feasibility. It will vary with technical advances and price changes. The criteria to assess discovery, recovery and commerciality of resources are well defined but vary between standards. Different definitions apply across the financial regulations that govern various stock markets where companies are listed. Interested readers can consult specific standards for the details of applied criteria.

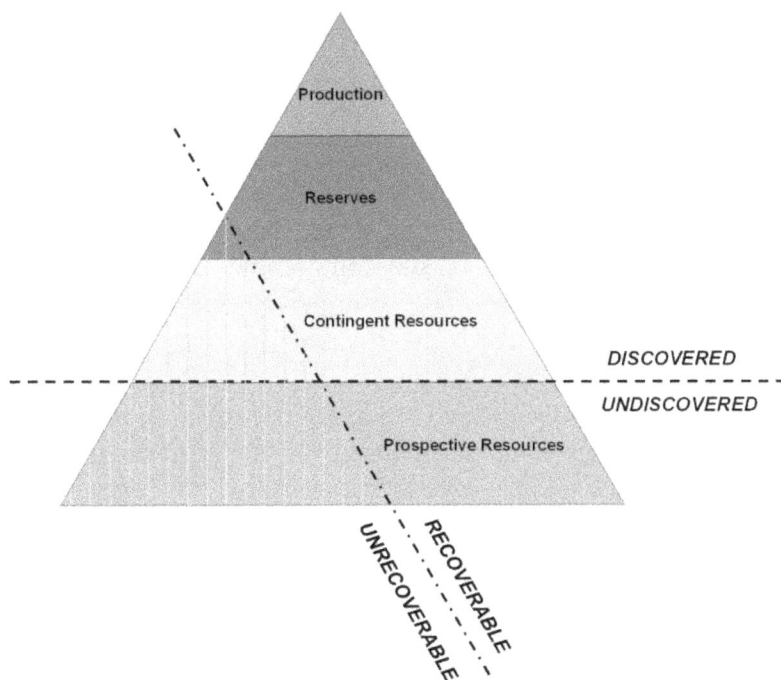

*Figure 3.1: Categorisation of fossil fuels resources*

Source: B.N. Asmar, Fossil Fuels in the Arab Worlds: Facts and Fiction, 2050 Consulting, 2010

The relationship between the above two criteria is illustrated in Figure 3.1, which places the reserves within the framework of

---

[4] Refer to Figure 3.2. Unrecoverable category is relative. It becomes smaller without commercial constraints as one could probably extract more of the resources then.

total resources. The figure also shows that three distinct portions of resources exist: *production, recoverable* resources and *unrecoverable* resources. It is worthwhile to note that the diagonal recoverable-unrecoverable interface can move both to the right or the left as a result of several interacting factors including fossil fuel price, extraction technology, political risks, and legislative conditions, to name a few.

In simplistic terms, the production refers to the cumulative quantities of fossil fuels already extracted, which are clearly recoverable. The remaining resources are either recoverable or unrecoverable. The sum of cumulative production and recoverable resources is often referred to as *ultimately recoverable resource* (URR), or *estimated ultimate recoverable* resource (EUR).

The remaining recoverable resources are divided into discovered and undiscovered.[5] In turn, the discovered resources are further divided into commercial and sub-commercial. Amalgamating these definitions together for simplicity, the recoverable resources can be described as divided into three distinct categories:

- **Reserves** - which are quantities of fossil fuels that are discovered, remaining in the ground awaiting production, recoverable and commercial. Reserves are further sub-categorised in accordance with the level of uncertainty associated with the potential recovery due to both feasibility and technology, with any of the two factors rendering the reserves proved, probable or possible.
- **Contingent Resources** - which are quantities of fossil fuels that are discovered, remaining in the ground awaiting production, potentially recoverable however they are not yet commercial. Similar to reserves, contingent resources are

---

[5] These undiscovered resources are often referred to as yet-to-find (YTF).

further sub-categorised in accordance with the level of uncertainty associated with the potential recovery due to both feasibility and technology, with any of the two factors rendering the reserves low, best or high estimates.

- **Prospective Resources** - which are quantities of fossil fuels that are undiscovered, remaining in the ground awaiting production, potentially recoverable and potentially commercial. Also similar to reserves and contingent resources, prospective resources are further sub-categorised in accordance with the level of uncertainty associated with the potential recovery due to both feasibility and technology, with any of the two factors rendering the reserves low, best or high estimates.

Figure 3.2 illustrates the above categories and sub-categories, where the dynamic and elastic nature of the boundaries defining the resources and reserves can be seen. Ultimately it illustrates the probability that a project will be developed and can reach commercial producing status. The figure presents the classification framework of fossil fuel resources as defined in the SPE-PRMS,[6] where the horizontal axis represents the range of uncertainty, which is a degree of geologic assurance and recovery efficiency, reflecting the range of estimated quantities potentially recoverable from an accumulation, and the vertical axis represents the chance of commerciality.

The unrecoverable resources are the portion of both discovered and undiscovered fossil fuels initially in place, which cannot be recovered at a given date due to geological, commercial accessibility or political constraints. However, a portion of these

---

[6] The Society of Petroleum Engineers (SPE) Oil and Gas Reserves Committee, made up of international oil and natural gas experts, partners with several industry related societies to provide publicly available resources for the consistent definition and estimation of hydrocarbon resources. As part of this work, SPE offers free documents including the Petroleum Resource Management System (PRMS), the PRMS Application Guidelines, as well as a map to other systems, and standards for reserves estimating and auditing.

quantities may become recoverable in the future.

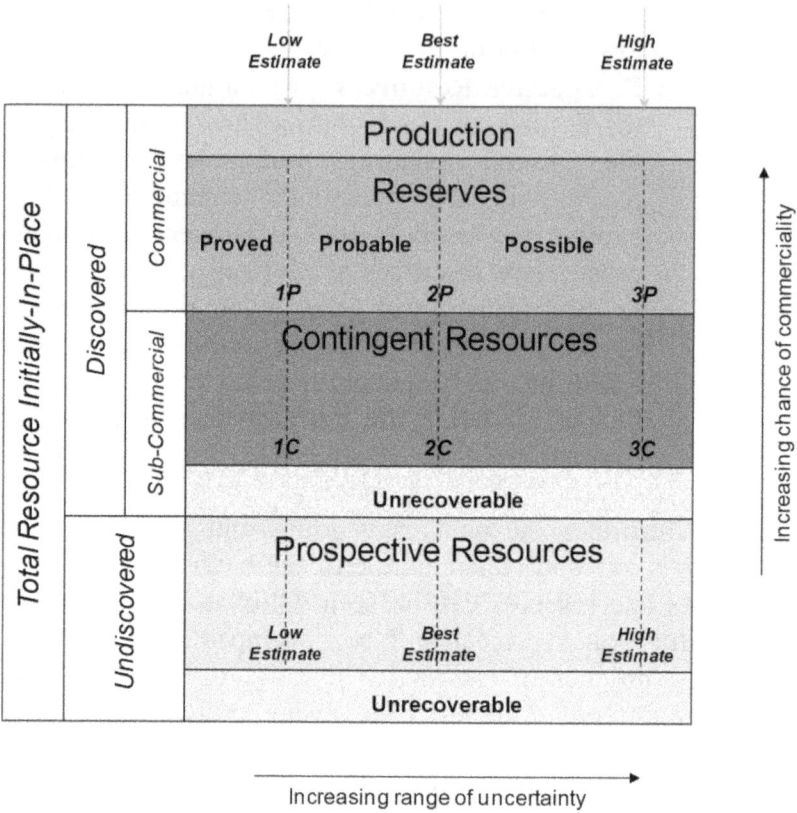

*Figure 3.2: Classification framework of fossil fuel resources*
Source: Modified from SPE-PRMS.

## 3.2   Oil Price

Oil pricing differs from most commodities or products, since
there are so many factors that play a role in setting its price.
These factors are often interlinked, yet they can contradict each
other, pulling the price in different directions. The price of oil is
not simply determined by forces of supply and demand, but
several other drivers including market sentiment towards both
current and future commercial contracts; speculation; geopolitical
conditions; specific political incidents; weather conditions;

environmental legislation, and the user/customer sentiment towards oil and natural gas, driven by ESG, which is eroding demand. As you can appreciate, forecasting oil price is an extremely complicated issue and one which is outside the scope of this book (see Box 1).

I mentioned earlier (see Section 1.5) that oil price collapsed spectacularly in March 2020 (as a result of the double whammy of oil supply glut and oil demand shrinkage).[7]

When the oil price collapse happened in March 2020, after demand shrank by over 20-25 million barrels,[8] no one expected that we would witness another collapse a month later, when US oil price crashed into negative territory, falling below zero for first time in history. Although this event was short lived, it shocked the public. It occurred when collapse in demand, left traders running out of storage facilities and trying to clear unwanted crude. The evaporation of oil demand, caused by Covid-19 pandemic, left the world with excess oil and not enough storage capacity. This effectively meant that producers were paying buyers to take crude oil off their hands. WTI oil price hit a record low of US$-40.32 per barrel, and closed in that memorable day US$-37.63, compared to US$18.27 on the previous day.[9,10,11] It is ironic that the Trump administration went from a stance of opposing OPEC[12] to begging it to rescue the market, officially backing moves by OPEC and Russia to cut production and pledging support for the industry.[13] It is also ironic though that

---

[7] The resulting carnage was unprecedented and price collapse surpassed the previous collapse, starting November 2014 and its aftermath when in January 2016 Brent oil price fell below US$28, which was a new low since 2003.

[8] https://www.hartenergy.com/exclusives/ogi-editor-chief-will-opec-cuts-matter-187277

[9] https://www.ft.com/content/a5292644-958d-4065-92e8-ace55d766654?shareType=nongift

[10] https://www.oedigital.com/news/477754-us-crude-futures-plunge-to-lowest-on-record?utm_source=AOGDigital-ENews-2020-04-21&utm_medium=email&utm_campaign=OEGDigital-ENews

[11] https://www.hydrocarbonprocessing.com/news/2020/04/traders-hightail-it-out-of-us-oil-contract-as-prices-reach-negative-40

[12] https://www.aljazeera.net/news/politics/2020/4/30/%D8%AA%D8%B1%D8%A7%D9%85%D8%A8-%D8%A7%D9%84%D8%B3%D8%B9%D9%88%D8%AF%D9%8A%D8%A9-%D8%AA%D9%87%D8%AF%D9%8A%D8%AF-%D8%AD%D8%B1%D8%A8-%D8%A7%D9%84%D9%86%D9%81%D8%B7-%D9%85%D8%AD%D9%85%D8%AF-%D8%A8%D9%86

[13] https://www.hartenergy.com/exclusives/will-us-shale-oil-rise-again-187247

this incredible shift occurred ten years, to the date, of the anniversary of Deepwater Horizon disaster in the Gulf of Mexico (See Section 2.3).

Other oil benchmarks also plunged badly that day, but did not break into negative territory,[14] with Brent, falling to the lowest price since 1999, hitting US$15.98 per barrel[15] and closing the day at US$19.33 per barrel[16] In fact the spot market prices for April 21st fell below US$10 per barrel.[17]

The price freefall halted after OPEC+ agreed to cut production and work towards helping to bring the market to a stabilised position. This was also assisted by many private producers who cut production, either voluntarily or were forced to, as their costs became untenable.

With price stabilising, creeping upwards into the 50sUS$ by the end of 2020, OPEC proved it still had influence on oil price. It proved speculators who argued that OPEC's influence was in only one direction i.e. flooding the market to destroy prices, wrong. It also proved it had influence in the other direction, as cutting production in a very low-price environment curtailed shale/tight oil production. The relatively low prices are preventing a significant restart of shale/tight oil production, where investors are more cautious and funding is increasingly tight.

This confluence of critical events showed how unprepared everyone was for "the unthinkable". For example, the Texas Energy Regulator considering oil production curtailments,[18] the American government asking OPEC and Russia to help stabilising oil markets, or considering stopping the oil tankers (carrying Saudi oil) from unloading in the US and adding to an

---

[14] https://www.hartenergy.com/news/why-all-eyes-will-be-expiry-brent-oil-futures-week-187320
[15] https://uk.reuters.com/article/uk-global-oil-idUKKCN224052
[16] https://www.statista.com/statistics/466293/lowest-crude-oil-prices-due-to-covid-19/
[17] https://www.eia.gov/dnav/pet/hist/rbrteD.htm
[18] https://www.hartenergy.com/news/texas-energy-regulator-vote-oil-production-curtailments-next-week-187279

American domestic glut spurred by the Covid-19 pandemic.[19] In addition, countries like Argentina planned to protect oil drilling by setting oil prices at a certain level.[20]

The crisis also brought into focus future price, ESG concerns and the level of investment available in the medium term. All these factors will play a pivotal role and, most probably, will be affecting the Middle East and the Arab world future position.

In 2016, most analysts issued and re-issued reviews of oil price forecasts, with the majority agreeing that US$100 by 2020 is highly unlikely and that hitting a low of US$20 is an actual possibility.[21,22]. These forecasts were proven correct, and at the time of writing, many analysts are in agreement that an oil price ceiling will be the norm for the years to come, with oil prices projected to stay generally low, with short-lived spiked occasionally.

It has long been believed that a major incident in the Middle East will result in a sudden price spike, the magnitude of which, depends on the severity of the event. I would suggest that, in fact, this is not true. Consider that, although tensions between Saudi Arabia and Iran in early 2016, hardly affected the price, it was speculated that a full-blown war between them may result in the spike exceeding US$500.[23,24] In fact, in September 2019, when the Houthis of North Yemen attacked Saudi Aramco facilities, - that resulted in taking out 5.7 million barrels a day of production capacity offline,[25] it caused a temporary price spike, of up to 15% for a day,[26] before subsiding quickly to below pre-attacks levels,

---

[19] https://www.worldoil.com/news/2020/4/30/saudis-43mmbbl-oil-flotilla-will-clog-more-than-us-storage-tanks
[20] https://www.worldoil.com//news/2020/5/8/argentina-s-plan-to-protect-drilling-by-setting-oil-prices-gains-momentum
[21] http://www.economist.com/news/finance-and-economics/21688446-why-oil-price-has-plunged-20-new-40
[22] http://oilpro.com/post/17197/don-t-be-surprised-if-oil-prices-hit-20?referer=1162633
[23] http://www.cityam.com/232262/oil-prices-could-hit-500-on-war-between-iran-and-saudi-arabia
[24] http://www.cityam.com/232262/oil-prices-could-hit-500-on-war-between-iran-and-saudi-arabia
[25] https://www.cnbc.com/2019/09/20/oil-drone-attack-damage-revealed-at-saudi-aramco-facility.html
[26] https://www.aljazeera.com/ajimpact/oil-prices-surge-attack-saudi-oil-facilities-

within a fortnight.[27] Similarly, the US assassination of General Soleimani of Iran in January 2020, caused only a minor blip in oil prices.

The events above demonstrate that major incidents, once perceived as significant risks to oil markets do not, in fact, have the impacts that were anticipated or feared for so long. We have witnessed a shift in oil price dynamics over the last few years. Historically oil was unique, as its supply was governed by lowest cost producer, unlike other commodities. However, it is moving to behave as a true commodity governed by marginal cost of supply.[28] Both short and long term factors will continue to interact, causing price volatility and external drivers, particularly environmental policies, will have an impact on its long term prospects.

According to Pulitzer Prize winning author Daniel Yergin (whose book "The Prize", provides a comprehensive history of oil and power), OPEC's days as economic force are "over".[29] With the organisation's failure to stop the decline in oil price, this has become increasingly evident. This view holds up in 2020, as we have seen that OPEC needed the cooperation of other oil producers, particularly Russia and even Trump's administration (overturning decades of American policy), to orchestrate a global oil supply management deal, to stop the oil price collapse. The events of the last few years, since the shale/tight oil revolution, has changed the correlation between oil price, oil production and politics fundamentally. While previously, oil price was the independent variable manipulating the dependent variables "production and politics", the reverse is now in place. Oil price has become the dependent variable, with producers and politics both influencing which way it is heading. Volatility will increase and oil price will be stuck in a narrow band, with its ceiling and floor much lower than even a few years ago.

190916003344259.html

[27] https://www.eia.gov/dnav/pet/hist/RBRTED.htm

[28] https://www.worldbank.org/content/dam/Worldbank/GEP/GEP2015a/pdfs/GEP2015a_chapter4_report_oil.pdf

[29] http://www.ft.com/cms/s/0/8a6e721c-fcb2-11e5-b3f6-11d5706b613b.html#axzz47WTcFjSs

Figure 3.3 shows the oil price trajectory since 1861, both in nominal and 2020 US dollars. It is obvious from the graph that price volatility is the norm, since we can easily see that price cycles occurred regularly. This demonstrates that the latest price decline of 2020 (Figure 3.4) is not a new phenomenon. What is different about this price decline is the reasons behind it, namely a double hit of supply glut and demand collapse.[30] While the supply glut has now receded, another spike is likely, and then structural changes to demand (i.e. competition from renewables / low carbon energy and government policies starting in OECD countries) will start to govern demand. This will be discussed in detail in the Section 3.4.

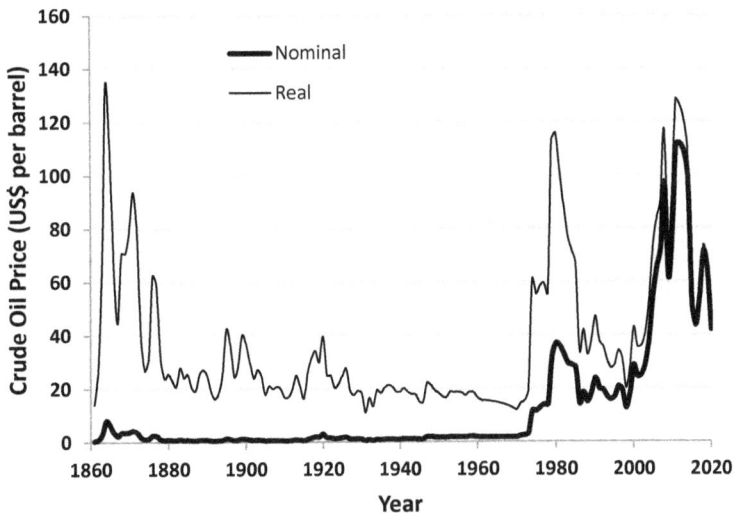

*Figure 3.3: Oil price trajectory*

Source: BP, EIA
Note 1: Consumer price index used from
(https://www.minneapolisfed.org/community/teaching-aids/cpi-calculator-information/consumer-price-index-1800).
Note 2: Prices are yearly averages.

---

[30] https://www.worldoil.com/magazine/2020/april-2020/features/world-oil-editorial-oil-market-meltdown-what-does-it-take-to-get-action

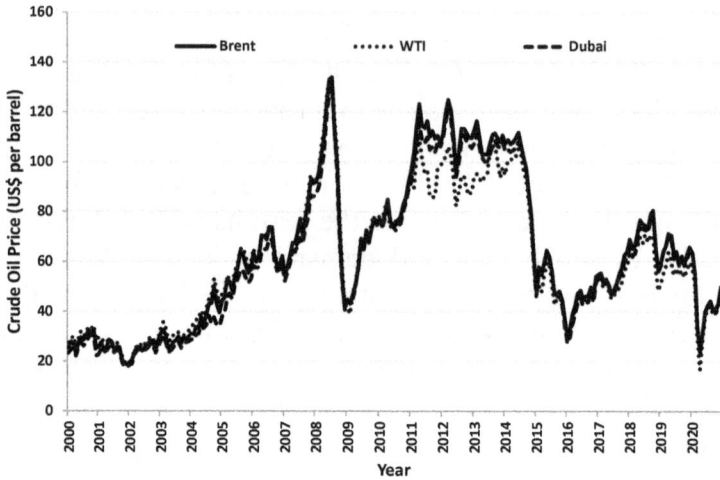

*Figure 3.4: Nominal crude oil price relative to 2000*

Source: IMF
Note 1: Prices are monthly averages.

## Box 1: Oil Pricing

The price of oil, often quoted in the media, refers usually to the price of a barrel of benchmark crude oil, in US$ per barrel. A benchmark is a type of crude that serves as a reference price for buyers and sellers. They are used to facilitate trade since there are many variations and grades of crude oil.

Three benchmarks are often used; these are West Texas Intermediate (WTI), Brent, and Dubai. Many other benchmarks are also traded including OPEC Reference Basket, Bonny Light, and Urals oil. In addition, benchmarks continue to evolve with new added periodically[31] including the Murban in the Middle East,[32] and the Platts American GulfCoast Select (AGS) in North America.[33]

---

[31] https://www.petroleum-economist.com/articles/markets/trends/2019/benchmarks-face-2020s-evolution
[32] https://www.cnbc.com/2019/11/11/middle-east-to-launch-a-new-oil-benchmark-to-rival-wti-and-brent.html
[33] https://www.cnbc.com/2020/06/26/oil-prices-sp-argus-launch-new-us-crude-benchmarks-to-rival-wti.html

There is a differential in the price of a barrel of oil, based on its grade, which is determined by either physical factors, such as its specific gravity (API) and viscosity; chemical composition, such as sulphur content; and geographic factors such as its location, in particular its export routes.

Heavier crudes (higher specific gravity) and sour (high sulphur) crudes are generally priced lower than lighter or sweeter crudes, because they are harder to transport and their refining requires additional equipment.

Producers and refiners often mix grades of oil into blends that meet defined specifications.

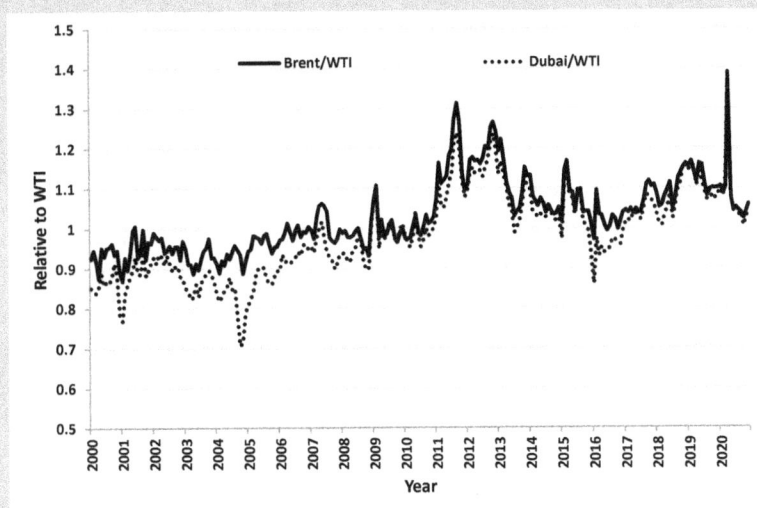

*Figure 3.5: Benchmark crude oil price differentiation (2000 – 2020)*

Source: IMF
Note 1: Prices are monthly averages.

Although superior blends often sell for higher prices, actual benchmark prices are determined freely based on market forces, and thus some benchmarks are sold at discounts sometimes.[34] See Figure 3.5.

---

[34] http://www.worldoil.com/news/2016/01/18/the-north-dakota-crude-oil-that-s-worth-less-than-nothing

Note that the prices quoted are often future prices rather than spot prices. But in the case of oil, the demand for immediate delivery versus future delivery is small, because the logistics of transporting oil to users make immediate spot exchanges rare. Thus, "futures contracts" are more common. An "oil futures contract" is an agreement to buy or sell a certain number of barrels, set amount of oil, at a predetermined price, on a predetermined date. These contracts can be monthly futures, monthly forwards, weekly Contract-for-Difference (CFD), or "dated" short-term assessed. A market is said to be in "contango" or "forwardation" when the futures price of oil is higher than the spot price, while the opposite is referred to as "backwardation".

## 3.3    Natural Gas Price

Unlike oil price, natural gas price is not commodified yet, although LNG pricing is heading this way.[35] Natural gas prices vary from region to region and how pricing is set (see Box 2).

In Section 1.5 I mentioned that natural gas prices were affected by oil price collapse of 2014 and in some regions, prices dropped by over 50%. As a result of the plummeting demand during the Coronavirus pandemic, oil price collapsed again, as did natural gas price and, in the summer of 2020, the spectre of prices turning negative, returned again.[36,37] Subsequently prices dropped to 25-year low,[38] which led to fears of inevitable shut-ins in production, due to the natural gas price plunge.[39,40] However,

---

[35] https://www.plattsinsight.com/insight/dawn-of-lng/

[36] https://www.worldoil.com//news/2020/6/3/natural-gas-prices-could-go-negative-on-global-oversupply

[37] Natural gas prices reached negative previously in certain areas, such as in Texas where producers had to pay for pipeline capacity, to be able to transport their natural gas away. See https://www.ft.com/content/ff4d72e9-65b6-3466-9002-f8b561191dde

[38] https://oilprice.com/Energy/Gas-Prices/Natural-Gas-Drops-To-25-Year-Low-As-Demand-Disintegrates.html

[39] https://oilprice.com/Energy/Natural-Gas/Natural-Gas-Price-Plunge-Could-Soon-Lead-To-Shut-Ins.html

[40] https://uk.reuters.com/article/uk-column-russell-lng-asia/column-spot-lng-the-worst-performing-

cold winters in Asia in 2020-21 reversed the LNG trend price, surprising the market and setting record high spot prices.[41,42]

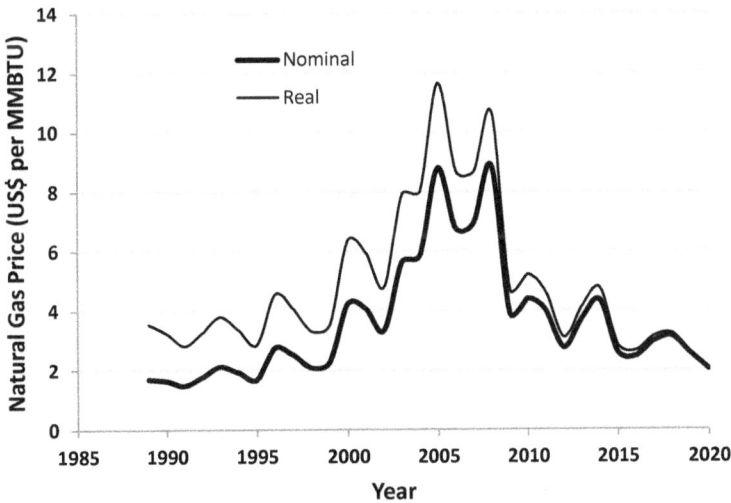

*Figure 3.6: Natural gas price trajectory – Henry Hub prices*

Source: BP, EIA
Note 1: Consumer price index used from
(https://www.minneapolisfed.org/community/teaching-aids/cpi-calculator-information/consumer-price-index-1800).

Figure 3.6 shows the natural gas price trajectory since 1989 both in nominal and 2020 US dollars. While the prices exhibit ups and downs, similar to oil price, the timing of these peaks and troughs differ. Figure 3.7 plots indexed oil and gas prices since 2000 at different regions, illustrating the regionality of the pricing. During the Covid-19 pandemic, regional price differences narrowed, and, in some instances, US LNG was the most expensive in the world[43] – a prospect never previously contemplated.

However, the correlation between oil and natural gas price is

energy-commodity-faces-more-price-pain-idUKKBN23B0PQ
[41] https://shippingwatch.com/carriers/article12680458.ece
[42] https://www.spglobal.com/platts/en/market-insights/latest-news/natural-gas/011321-factbox-asian-spot-lng-prices-hit-record-highs-on-supply-glitches-demand-spike
[43] https://www.hartenergy.com/news/lng-buyers-cancel-cargoes-us-natgas-becomes-most-expensive-world-187467

diverging and the future price movements are expected to separate.

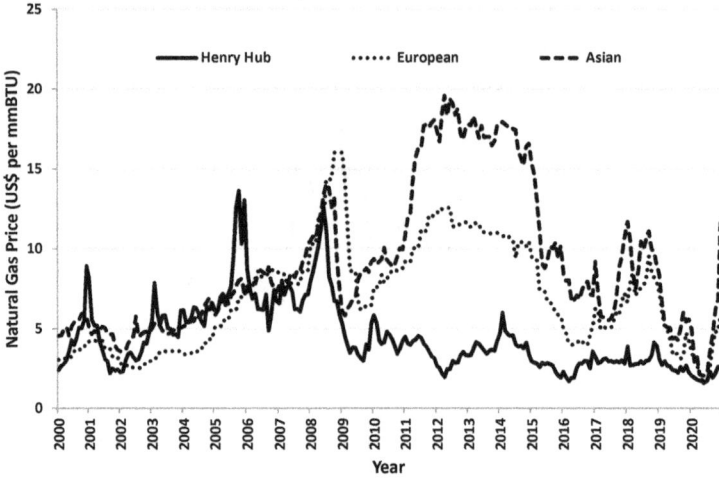

*Figure 3.7: Nominal natural gas price relative to 2000*

Source: IMF
Note 1: Prices are monthly averages.
Note 2: Year 2000 equals 100.

## Box 2: Natural Gas Pricing

Unlike oil, natural gas price differs from region to region. It is always priced based on its energy content and is often measured in US$ per mmBTU.[44]

In North America, it is priced as a commodity, while in Europe and Asia, the prices are often linked to the price of crude oil and/or petroleum products. Spot pricing is being gradually adopted in all regions.[45]

Since the composition of natural gas is often homogenous, its benchmarks are often geographic hubs and are affected by the trade routes. The most important is the Henry Hub in North

---

[44] One million British Thermal Units
[45] https://www.eia.gov/todayinenergy/detail.cfm?id=23132

America, the National Balancing Point (NBP) in the UK, and the Title Transfer Facility (TTF) in the Netherlands.

Actual benchmark prices are determined freely, based on market forces so some benchmarks are occasionally sold at discounts. See Figure 3.8.

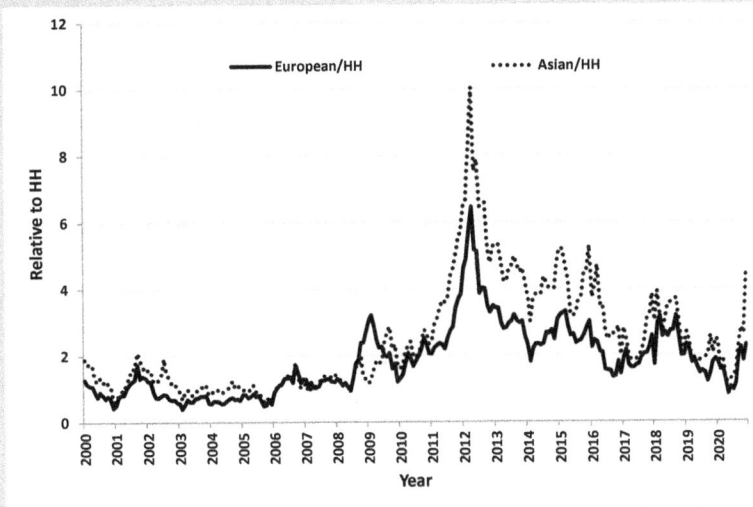

*Figure 3.8: Benchmark natural gas price differentiation (2000 – 2020)*

Source: IMF
Note 1: Prices are monthly averages.

## 3.4   Oil and Natural Gas Markets: Supply Glut or Demand Peak?

Following the oil price collapse of 2014, many analysts blamed increasing oil and natural gas supplies as the main reasons that prices dropped and remained depressed. It is important to point out that the markets are symbiotic, especially in terms of production. A substantial amount of natural gas is produced, as associated natural gas, when producing oil equally, large amounts of natural gas liquids (NGLs) are produced (as a by-product) when natural gas is produced. This symbiosis is an important factor affecting both supply and demand markets, as well as the prices of both oil and natural gas.

In 2018 there was some measure of optimism that the oil market might have turned a corner and that an oil price recovery was underway. That optimism proved to be short-lived. Events in the first quarter of 2020 turned the market on its head. Saudi Arabia triggered an oil price war by increasing its oil production to flood the oil markets. This led to OPEC's production hitting a thirty-year high.[46] The timing of this action was particularly unfortunate, coinciding, as it did, with the Covid-19 pandemic, which caused an unprecedented reduction in demand. The combination of the two events led to a dramatic oil price collapse in March and April 2020 (See Section 3.2). Despite OPEC+ taking action to stabilise the oil markets, the increasing belief amongst industry experts is that prices will recover only to 2014 prices in the short or medium term. Many experts believe that prices will remain in a narrow band, considerably below 2014 price levels for the foreseeable future.

Ironically, in March 2020, Saudi Arabia indicated that it did not want to manage the oil market, but its earlier actions turned a demand crisis into a supply catastrophe for the oil industry. In April 2020, it returned to managing the market and coordinated with allies, to compensate by cutting production, more than required, in a frantic effort to stabilise oil price. This involvement continues (at the time of writing), with unilateral production cuts as part of the Saudi intervention, as seen in early 2021, of a one million unilateral oil production cut, in an effort to stop oil prices from dropping again.[47]

Meanwhile, the natural gas market continued to grow. It had been promoted as the bridge fuel, to slash the world's oil dependence, while buying time to develop alternative energy sources to reduce greenhouse gas emissions. During the period, 2015 to 2019, there was a flurry of natural gas projects developing several LNG export terminals, especially in Australia and the US. This

---

[46] https://www.worldoil.com//news/2020/5/1/opec-output-hit-30-year-high-during-the-saudi-russia-price-war
[47] https://www.wsj.com/articles/saudi-arabia-russia-reach-compromise-on-opec-plus-production-plan-11609857544

additional capacity created excess supply which, coupled with low demand due to Covid-19 crisis, exerted pressure on natural gas prices. In early 2020 this led to record low prices and the disappearance of differentiation of regional prices (for a limited period). However, with the reduction in oil production, associated natural gas volumes will be cut, resulting in a tightening of the market and projected price increases in the longer-term. Nonetheless, periodic spikes in price will occur due to certain conditions, such as cold weather or supply bottlenecks.

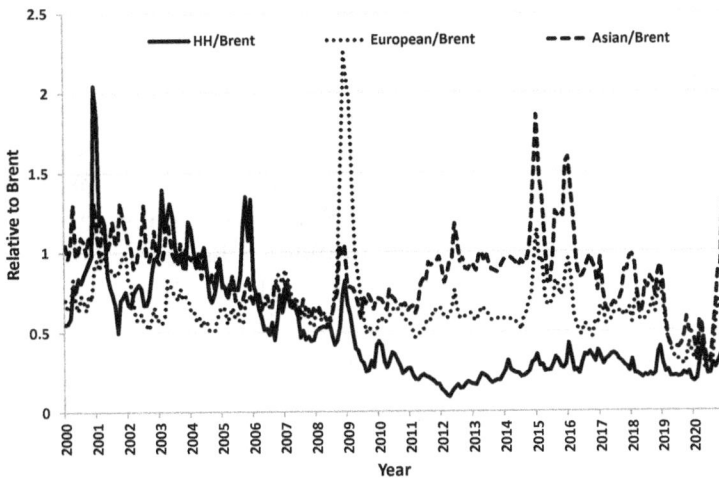

*Figure 3.9: Relative nominal natural gas to crude oil price (2000 – 2016)*

Source: IMF
Note 1: Prices are monthly averages.

It is important to differentiate between oil and natural gas. Figure 3.9 shows the natural gas price as a function of oil price. As stated earlier, the two are now diverging, especially in North America. This means that, relatively speaking, there may be a stronger case to develop natural gas projects in tough fiscal climates, where gas may be more competitive in situations relative to oil, assuming fiscal terms do not change. On the other hand, one can conclude from the above figure that if oil price remains low then the fiscal benefit of natural gas diminishes in certain regions.

Consequently, the link of natural gas price to oil price, which had already been broken in any instance, will be hard to restore due to oil price volatility, variable demand patterns, multiple players, political situation and economic liberalisation processes.

Thus, examining the new reality of energy markets following the seismic hit in 2020, one can make the following observations:

1.  Prior to Covid-19 pandemic, oil supply has increased year on year since 2010, while at the same time although oil demand also increased, its growth rate was lower. (See Figure 3.10.) Both supply and demand crashed in 2020 leading many to suggest that peak demand has been brought closer.[48] Debate is raging as to whether oil demand will ever return to its 2019 levels or have we already hit peak oil demand? Some companies, including BP[49,50] Shell,[51] Total[52] and Equinor,[53] believe we reached the peak or we are about to reach it. Others, such as Exxon, do not subscribe to this notion yet. Similarly, opinions published by consultants (e.g. IHS Markit, Wood Mackenzie),[54] think tanks and organisations differ, with all agreeing that the Covid-19 pandemic has brought the peak time closer.[55]

    Natural gas supply has increased, year on year, since 2010, however, its demand growth rate is increasing at a similar rate to the supply rate. See Figure 3.11. Natural

---

[48] https://uk.reuters.com/article/us-global-oil-vitol/top-trader-vitol-says-virus-might-bring-peak-oil-demand-much-quicker-idUKKBN22H1P3
[49] https://www.hartenergy.com/exclusives/bp-report-global-energy-consumption-rises-growth-rate-slips-188155
[50] https://oilprice.com/Energy/Crude-Oil/BP-Boss-We-May-Have-Already-Hit-Peak-Oil-Demand.html
[51] https://www.nytimes.com/2021/02/12/business/dealbook/shell-peak-oil.html
[52] https://www.worldoil.com/news/2020/9/29/total-joins-bp-in-projecting-an-oil-demand-peak-around-2030
[53] https://www.reuters.com/business/energy/pandemic-brings-forward-predictions-peak-oil-demand-2021-09-28/
[54] https://oilprice.com/ 15/05/2020
[55] https://www.worldoil.com/news/2021/1/14/mckinsey-projects-a-2029-oil-demand-peak-accelerated-by-covid-19

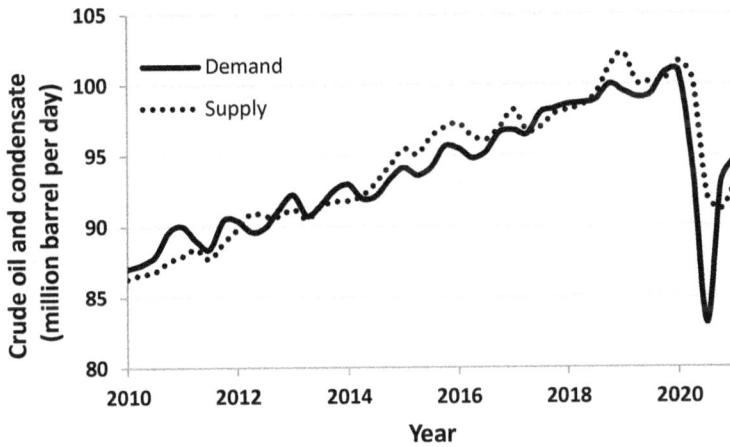

*Figure 3.10: Global oil supply and demand (2010 – 2020)*

Source: IEA
Note 1: Number includes crude oil, condensate, and other liquids.
Note 2: Quarterly data

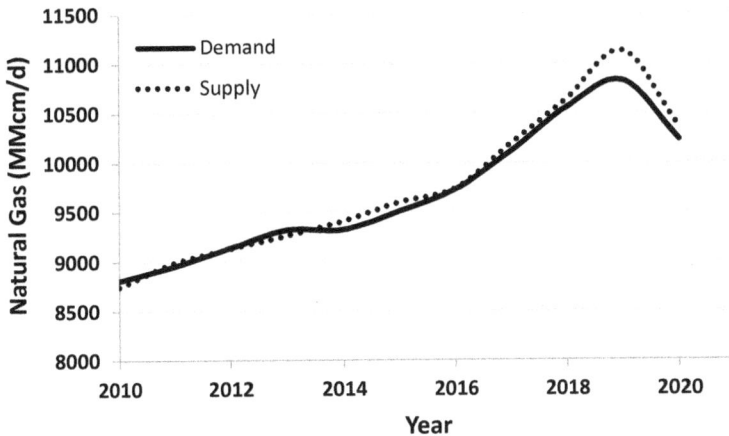

*Figure 3.11:Global natural gas supply and demand (2010 – 2020)*

Source: IEA
Note 1: Yearly data

gas supply and demand crashed in 2020, as with oil.
Likewise, there is debate, not about whether demand will
ever recover to its 2019 levels, but rather the timing of

peak demand.[56,57]. In the case of natural gas, it is forecast that demand will continue to grow in the shorter term, although some do suggest that peak natural gas demand, will occur only a few years after peak oil demand.[58]

2. Conventional oil and natural gas discoveries have fallen to their lowest level in 70 years.[59] The reason for this decline is twofold. First, the aftermath of 2014 oil price collapse caused companies to significantly reduce their exploration budgets. Second, financing from investors dried up as many favoured short-term returns of shale/tight oil, that provided more flexibility in response to market volatility. Furthermore, exploration success rates have been reducing, being at their lowest level since 1950. The average discovery size also decreasing considerably.

However, despite the low level of discoveries, prior to the Covid-19 pandemic, oil and natural gas reserves were in fact increasing. This was due to rapid advances in technology, especially advances in seismic technology and drilling, allowing increased recovery from existing reservoirs and enabling access to new reservoirs such as the Arctic or deepwater frontier fields.[60] In addition, new basins[61] have established themselves as new promising areas such as offshore Guyana, offshore East Africa and the Eastern Mediterranean region. In the past five years the unconventional shale/tight oil and natural gas revolution, is the clearest proof of the technology impacts, resulting in massive production increase. As with many

---

[56] https://www.petroleum-economist.com/articles/midstream-downstream/lng/2020/gas-and-lng-brace-for-tougher-times

[57] https://www.worldoil.com//news/2020/6/10/decline-in-natural-gas-demand-may-take-years-to-reset

[58] https://www.worldoil.com/news/2021/1/14/mckinsey-projects-a-2029-oil-demand-peak-accelerated-by-covid-19

[59] https://www.worldoil.com/news/2019/10/1/conventional-discoveries-fall-to-lowest-levels-in-70-years-and-a-major-rebound-isn-t-likely-says-ihs-markit

[60] Basel Asmar, Subsea Processing: A game changer waiting in the wings? IHS Special Report, 2016

[61] A basin can contain several reservoirs.

new technologies, people can be slow to adapt to change. These new technologies have allayed the fears of those who feared[62] for the future production potential of oil and natural gas, quickly proving them to be a resilient resource class. In fact, the pace of development took many by surprise. The pandemic aftermath may prove to be a pivotal point, as the persistence of low oil prices will possibly make some reserves uneconomical, leading to their reclassification as resources again, as they will become economically unfeasible. It may also put other assets on hold as companies will find it hard to justify developing them, e.g. many Arctic resources.[63]

Although the pandemic may not be the main cause, it has amplified the shift in ESG concerns. So too has the track record of poor returns, which influence US (and other) producers' capital discipline and short-termism. These factors, alongside price signals, will determine the level of investment, and hence production growth, from US shale, as well as new frontiers like the Arctic.

Taking all these factors into account, we need to accept that some oil and natural gas reserves may be destined to remain in the ground and never be produced.

3. The oil price collapse of 2014 forced the oil and natural gas operators to fundamentally change the way they work, by adopting and embracing technology, it enabled them to perform more efficiently. The improvements made then, benefitted them in 2020, helping them cope with the effects of the second oil price collapse and placing them in better position to recover, once this latest downturn ends.

4. There is an International commitment to cut usage of fossil fuels mostly targeting coal and oil. Recently, noises emission concerns have been raised about natural gas, by

---

[62] http://viableopposition.blogspot.ca/2014/12/overstated-tight-oil-reserves-and-false.html
[63] https://www.bloomberg.com/news/articles/2020-04-24/wall-street-is-bending-to-pressure-to-halt-arctic-drilling-loans

people who no longer recognise natural gas as part of the Green Policy pathway. In the last few years, due to increased public awareness of climate change issues and tightening environmental legislation in many countries in the world, growth in oil demand has slowed down. In order to comply with the new regulations on emissions, and the squeeze in capital spending on "dirty" fossil fuel projects, producers are utilising advances in energy efficiency technology to curtail consumption and reduce emissions. Recently there has been a clear trend in developed countries that the public is using less energy (per capita) for transport or heating,[64] and reaping economic benefits as a result. This "win-win" has meant more money in their pockets with less oil and natural gas demand growth overall.

5. The costs of alternative energies are falling rapidly, transforming them into real competitors to fossil fuels. This is due to technology advances and governmental policies providing incentives and subsidies, which drives the share of fossil fuels demand lower in percentage terms, (although until the Covid-19 pandemic, fossil fuel demand was still rising in absolute terms).

While many of the above observations continue to be true, parts of the story have changed radically as a result of the pandemic and in some instances, the change will be irreversible. The outlook for both the oil and natural gas markets is challenged. Oil has certainly joined coal as a "dirty fuel" even in the eyes of some major oil producing companies, while natural gas is no longer widely accepted as a bridge fuel. These changes have made funding for future oil and natural gas projects is more challenging.[65]

Prior to the pandemic, the shale/tight oil and natural gas revolution killed off the original "Peak Oil" theory, that

---

[64] http://www.eia.gov/forecasts/ieo/world.cfm
[65] https://www.oedigital.com/news/478450-japanese-bank-puts-oil-sands-arctic-drilling-on-restricted-list?utm_source=AOGDigital-ENews-2020-05-13

advocated reserves peak and supply crunch, completely. In the years leading up to the Covid-19 pandemic, shale / tight oil was a major contributor to production growth. Today, it is also true that reserves may not be a critical limitation either - the world is awash with the black stuff. The oil and natural gas industry has proven that there are more resources that can be developed, subject to the application of technology, assuming the price is right. Technology has proved vital in increasing resources and reserves while the oil and natural gas industry has continued to ensure its viability with responsive innovation. The more expensive oil and natural gas becomes, the more reserves will be added and reclassified, moving from resources.

Ironically, resource abundance is offset by constrained investment and the de-emphasis of fossil fuel development, in favour of renewables. If the situation does not change, there is a risk of a significant supply crunch and price spike in the next few years.

Examining the oil and natural gas of the US over the last decade, demonstrates the transformation that turned it into the world's largest oil and natural gas producer by the end of 2019. See Figures 3.12 and 3.13.

However, as a result of the 2020 oil price collapse, the companies involved in the tight/shale oil industry suffered such colossal decline, that some observers described it as "a rout".[66] To many, it appeared to be a swift and brutal end to the tight/shale oil revolution.[67] By the end of 2020 some growth returned, but at a slower pace. Many analysts believe that unconventional oil production has already peaked,[68] while others project it will take years to recover,[69] if ever.

---

[66] https://www.ft.com/content/a5292644-958d-4065-92e8-ace55d766654
[67] https://www.worldoil.com//news/2020/4/22/oil-rout-looks-to-break-shale-s-global-oil-dominanc
[68] https://www.wsj.com/articles/coronavirus-threatens-to-hobble-the-u-s-shale-oil-boom-for-years-11590312601
[69] https://www.worldoil.com//news/2020/6/24/shale-oil-production-may-take-years-to-recover-despite-a-short-term-uptick?id=1774035

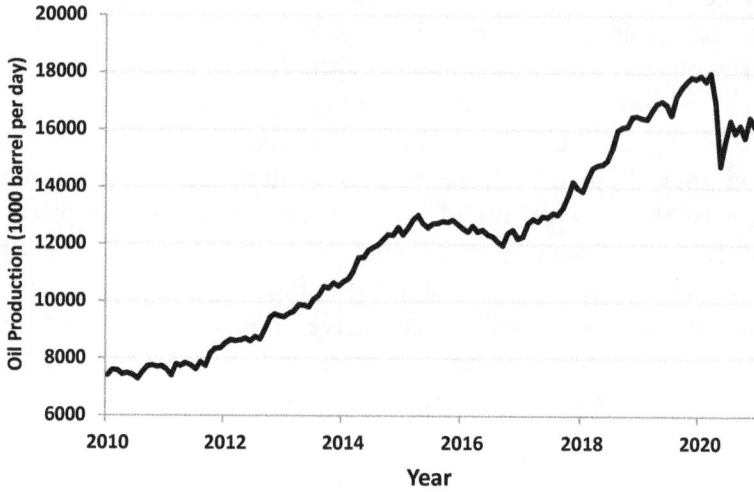

*Figure 3.12: United States crude oil and condensate production (2010 – 2020)*

Source: EIA
Note 1: Number includes crude oil, condensate, and NGL

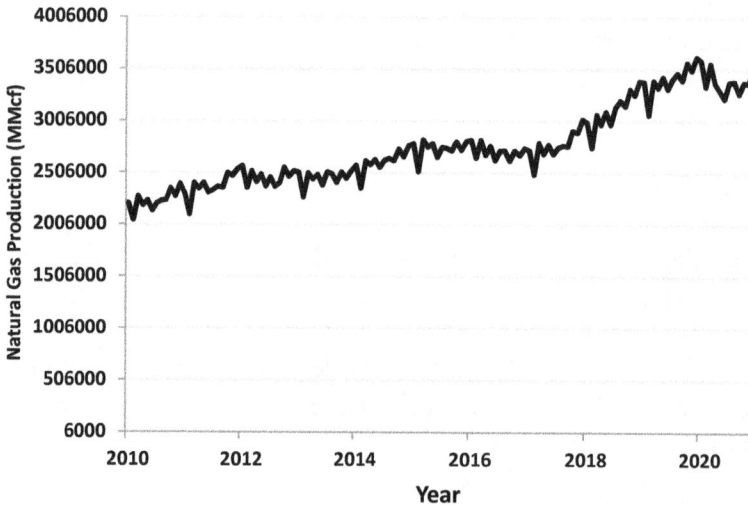

*Figure 3.13: United States natural gas production (2010 – 2020)*

Source: EIA
Note 1: the number reported is for gross withdrawals, which are the full well-stream volume, including all natural gas plant liquids and all nonhydrocarbon gases, but excluding lease condensate. Also includes amounts delivered as royalty payments or consumed in field operations

In the last five years, technical advancements resulted in expanding the volume of utilised endowment of "originally-in-place" fossil fuels. These additions to resources and reserves exceeded consumption in the same period, so the absolute volume, the fraction of total oil and natural gas resources remaining in 2020, is greater than in 2015, despite the additional five years actual consumption. This contributed to the R/P (reserves to production) ratio staying relatively unchanged and remaining stable.

Ironically, the collapse in oil price of 2014 benefited the industry in certain aspects, as it proved to be a catalyst improving efficiency and decreasing production costs. These new approaches caused structural changes to project execution. the combined efficiency gains and technology advances led to significant cost reductions. The combination of many incremental steps, improving production and processes, made a big difference on aggregate. This is not unprecedented, as history tends to repeat itself. For example, the 1986 oil slump led to several drilling breakthroughs, that formed the basis for adopting horizontal drilling globally. This was at the heart of the tight/shale oil and natural gas revolution. The resultant gains enabled the development of more resources, leading to increased production.

Thus, from a technical point of view, pre-Covid-19, it was widely accepted that oil and natural gas supply and demand will continue to grow in the short and medium term, despite environmental pressures. Unless radical existential changes occur - such as in people's attitude to energy consumption and fossil fuels, particularly the tightening of environmental regulation, or increased taxation to curb oil and natural gas demand. Otherwise, the status quo could prevail, with the benefits of cost efficiencies from further industry restructuring, continued rapid technology development which enables both enhancements to supply, and reduced consumption (hence, reducing demand) through more efficiencies combustion engines, for example.

In addition, pre-pandemic, it was expected that many countries would be seeking to mimic the US shale gas and tight oil revolution.[70] Technically, this can be achieved worldwide. Advances in fracking technology continue to progress and can in principle be applied elsewhere. Progress was already being made in China, which is currently producing commercial natural gas from its unconventional plays. Argentina, along with various international partners, was also developing its oil and natural gas production in its Vaca Muerta Basin. Other countries are slowly following in the footsteps, including Russia, Algeria, Saudi Arabia, and UAE. For example, Saudi Arabia has started the US$110 billion Jafurah project[71] to tap its unconventional natural gas reserves. The Kingdom is benefiting from the decline in activity in North America's unconventional plays. It has actively recruited entire crews and, utilising from the expertise of others, starting its own production, from a higher point on the learning curve. This is classic in the industry, where fast followers often reap the benefits from first movers. The UAE has started developing its shale gas reserves at Ruwais Diab, with collaboration between ADNOC and the French company Total (now TotalEnergies).[72] Nevertheless, the uncertain outlook of oil and natural gas demand market, with few exceptions, e.g. China, Argentina and Russia, will certainly slow down many of the unconventional development projects for now.

In these Covid-19 pandemic times, projections of oil and natural gas demand have changed drastically. Many analysts, organisations and companies believe that oil demand has peaked, or will peak within the next five to ten years. The potential shale/tight oil industry may be diminished in some regions and disappearing in others, before it has even had an opportunity to get off the ground. At the moment, the shale/tight oil (and to a lesser extent natural gas) bubble appears to have burst. But this is not written in stone, as another revival could happen, if geopolitics change in the Middle East or if oil price defies analysts forecasts and rises sharply again.

---

[70] http://www.ft.com/cms/s/0/7d892f7e-3a07-11e5-bbd1-b37bc06f590c.html#axzz3yut9GJPf
[71] https://www.arabnews.com/node/1631841/business-economy
[72] https://www.worldoil.com/news/2020/11/11/adnoc-and-total-deliver-first-unconventional-gas-from-the-uae

Covid-19 has indeed changed the narrative. What was highly publicised by the Trump administration as "American Energy Dominance"[73,74] is no longer valid. The events that followed the 2020 oil price collapse, where rumoured threats (by Trump's administration) to withdraw military support to Saudi Arabia as an extortion tactic unless it buys American weaponry as a form of compensation, and any rhetoric about enacting the NOPEC legislation was ditched.[75] The Biden administration is considering its own mandatory oil cuts.[76] That it is effectively begging OPEC+ to interfere in the market, is a truly remarkable and unprecedented event,[77] which until March 2020 was totally both unexpected and unthinkable.

I would suggest that writing off fossil fuels from the future energy picture is premature. Despite the wishful thinking and many opinion pieces, writing about the eventual demise of the oil and natural gas industry, it is important to remember that contrasting forces exist that can alter this trajectory. The versatility of oil as an energy source is unparalleled and, at present, it will be hard to replace it easily or cheaply.

Just consider that the world's increased population is creating more energy demand and, while the demand percentage growth will be slowing[78] (e.g. according to the IEA from 1% for the period between 2010 and 2020 to 0.3 % for the period 2020 to 2040), actual demand by volume will increase. The increase in demand is compounded by growing urbanisation, leading to more energy consumption.

Figure 3.14 illustrates that oil demand will remain significant and that natural gas demand will continue growing, according to IEA base scenario. When we compare the projections of 2015 to, both

---

[73] https://www.hartenergy.com/exclusives/role-oil-and-gas-industry-energy-transition-186597
[74] https://www.ft.com/content/a5292644-958d-4065-92e8-ace55d766654
[75] https://www.hartenergy.com/news/sources-say-trump-told-saudis-cut-oil-supply-or-lose-military-support-187331
[76] https://www.hartenergy.com/news/texas-energy-regulator-drops-plan-oil-production-cuts-187381
[77] https://www.hartenergy.com/exclusives/will-us-shale-oil-rise-again-187247
[78] https://www.hartenergy.com/exclusives/bp-report-global-energy-consumption-rises-growth-rate-slips-188155

pre-Covid and post-Covid-19, projections of 2020, it highlights the declining fortunes of oil in that relatively short period. Despite this decline, oil will remain significant for years to come. Other predictions are more aggressive and forecast steeper oil decline. These are discussed in Section 3.6 (energy transition).

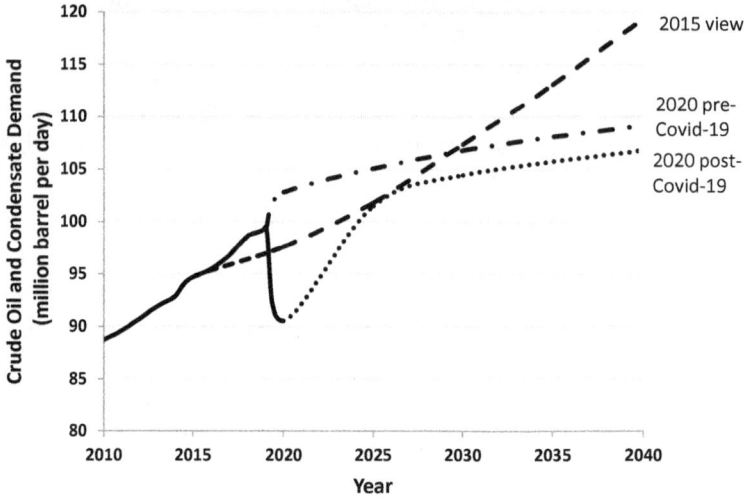

*Figure 3.14: Global oil projected demand (2015 – 2040)*

Source: IEA
Note 1: dashed and dotted lines are projected after the year of actual data point in the corresponding report.

Another consequence of the 2014 oil price collapse is the change in, oil and gas economics which has resulted in deferring various projects. In previous oil price collapse cycles, it was often argued that the deferral of projects would eventually lead to a squeeze in supply,[79] which would wipe out any oil glut. At that inflection point, oil price will increase,[80] followed by production costs increase and another upward cycle. The reality of previous cycles does not support this. What we witnessed was that lower oil price triggered innovation and ingenuity, which led to new ways for the industry to cut costs, survive and then re-emerge, triggering a new upward cycle. Those supply crunch fears have not

---

[79] http://www.eia.gov/finance/markets/reports_presentations/eia_what_drives_crude_oil_prices.pdf
[80] At the time of writing, it appears we have reached the inflection point.

materialised previously, and the feared supply crunch never happened – at least, not yet.

However, it is important to realise that there is not an on/off switch for oil production from oil fields. Stopping and starting production is not simple, and good forward planning is necessary to maintain availability of smooth supply. Several factors need to be considered when resuming production, from an oil field, following a halt in its production. These include technical and logistical factors, such as field maturity, restart complexity[81] and storage availability. In addition, financial considerations such as oil price, operating margins and the financial health of the companies. Regulatory and contractual conditions also need to be taken into consideration.[82]

In 2020 the Covid-19 pandemic delivered another blow to oil demand. The decline in demand certainly brought forward the projected oil demand peak and accelerated its downward trajectory. However, it did not bring it to the much-anticipated end – there are many years of oil demand ahead of us. As seen in Figure 3.14, the IEA oil demand outlook does not show peak demand before 2040, but it does illustrate that the demand volume is reduced and shows how the pandemic led to a structural downgrade in demand. It is important to stress that the main driver changing the outlook between 2015 and 2020 is the energy transition and, inevitably the peak demand is being brought forward.

Producers understand that eventual decline will be inevitable and will become irreversible. Therefore, they are pursuing different strategies to prepare for the next phase of the age of fossil fuel. Many are being forced to reduce investment and write off huge amounts of asset value, due to lower oil and natural gas prices.[83,84]

---

[81] https://www.worldoil.com//news/2020/5/8/operators-prepare-for-technical-and-fiscal-challenges-restarting-shut-in-wells
[82] https://www.worldoil.com//news/2020/5/8/crashing-oil-demand-drives-a-17-mmbpd-global-output-cut-in-q2
[83] https://www.oedigital.com/news/478517-brazil-s-petrobras-warns-economy-has-changed-forever-books-11-2b-impairment?utm_source=AOGDigital-ENews-2020-05-15

In this aspect, there is distinct behavioural difference between national oil companies (NOCs), whose policies are dictated by the political wills and policies of their owners, and international oil companies (IOCs), who are accountable for their shareholders. The latter companies are pursuing two distinct paths, where, broadly speaking, the European companies are on an accelerating energy transition pathway, whereas the American companies are more passive in that regard,[85]. In addition, European companies broke with tradition by cutting dividends to investors – an absolute "No" for American companies. Some NOCs are becoming hybrid international companies, which can be labelled as INOCs, where they are selling a stake in themselves, or are entering into numerous joint ventures. Companies like ADNOC and Qatar Petroleum are moving in this direction. Even the national Venezuelan oil company PDVSA is moving in this direction.[86]

In the short and medium term, competition between fossil fuel types, over market share, will intensify. As seen in the previous sections 3.2 and 3.3, the initial collapse in oil price was accompanied by lowering natural gas prices. Oil and natural gas prices diverged later, due to the loss of associated natural gas production and a cold winter increasing natural gas demand. These low prices experienced in the first half of 2020, put further pressure on coal prices, which lost approximately 20% in the same period, before recovering later (see Figure 3.15). Coal is being squeezed by the tightening of environmental legislation and the fall in prices of all other energy forms. The relative price movements are currently very dynamic, and analysts have started wondering if we have arrived at a point of displacement, i.e. where one fuel (e.g. natural gas)[87] totally replaces another (e.g. oil or coal).

---

[84] https://www.bcg.com/en-gb/publications/2020/oil-and-gas-investment-during-the-covid-era
[85] Note that through 2021 it is evident that American companies are beginning to adopt more progressive Energy Transition efforts, not least due to shareholder influences.
[86] https://www.worldoil.com//news/2020/4/29/pdvsa-considers-dramatic-changes-to-revive-venezuelan-oil-production
[87] https://www.naturalgasworld.com/editorial-getting-greener-no-longer-costs-the-earth-lng-condensed-79387

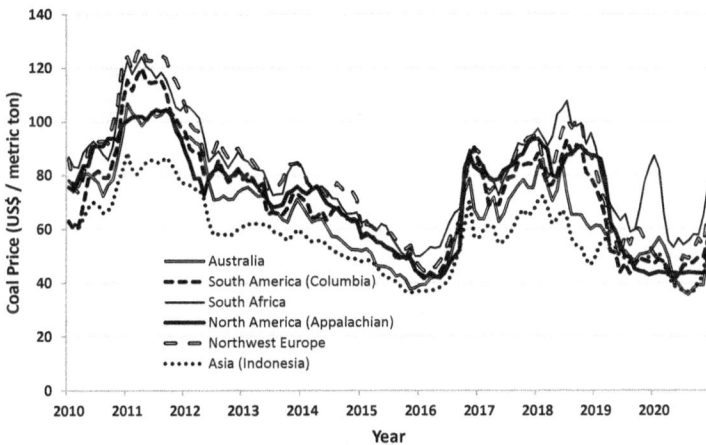

*Figure 3.15: Nominal coal price (2010 – 2020)*

Source: World Bank; IHS McCloskey; Quandl
Note 1: Prices are monthly averages.

We will be seeing a structural change in oil consumption, fundamentally altering oil demand and people's attitudes,[88] with more people working remotely, international travel being reduced and supply chains brought closer to home. Recovery will depend on changing levels of mobility and resumption of travel.[89] Years ago, the late Minister Yamani of Saudi Arabia said the oil age will end before we run out of oil. His prediction appears to be coming true.

## 3.5   Costs Second Reset – Lower for Longer is the New Norm Now

Following the oil price collapse in 2014, oil and natural gas producers had to make structural changes in the way they operate in order to survive and stay competitive. They had to go back to the drawing board and rethink their strategies and outlooks. Many projects were deferred, redesigned, and processes were performed more efficiently to cut both capex and opex.

---

[88] https://www.worldoil.com//news/2020/4/30/oil-continues-climb-on-signs-of-a-coming-demand-recovery
[89] https://www.hartenergy.com/exclusives/want-oil-price-recovery-hit-road-187282

The actions implemented in 2015 and 2016 resulted in significant cost savings, where overall costs were reduced by over 40% generally.[90,91,92] The biggest savings (in magnitude) were achieved in offshore projects, where their survival depended upon fundamental changes to the way they were managed and executed.

Following the rapid cost reset in those two years, the market appeared to have been stabilised. Cyclical (and inflationary) costs rose, while structural costs decreased, thus dampening inflationary increases. See Box 3. However, the second oil price collapse of 2020 triggered another cycle of cost reduction in an already difficult market conditions, which were not yet fully recovered from the previous cycle. A key difference, from this 2020 collapse was that all sectors were suffering (as the effects of the pandemic hit them all), while previously a downturn in one sector benefited other sectors.[93] In the lower oil environment of 2020 and early 2021, there was little room to cut costs more, as the margins are already tight. Here again, structural cost reductions, including simplification and the use of repeat designs, will be vital for the industry to survive. The industry needs to adapt to be able to operate profitably in the future when lower oil prices could be the new norm.

Figures 3.16a and 3.16b show the costs trajectory. The capital and cost upstream indices, published by IHS Markit, track the cyclical costs, clearly illustrating the two cost cycles of 2014 and 2020, as well as a previous cycle of 2008. The structural costs are tracked by the upstream innovation index, showing cost reductions due to gains in efficiency, technology and design changes. The tracking only goes back to 2014, following the previous oil price collapse. A future cost recovery is forecasted to be gradual and slow.

---

[90] Basel Asmar, Pritesh Patel, Susan Farrell, Raoul LeBlanc The great cost reset: The new competitive dynamic in US unconventionals, global deep water, and the Middle East, CERAWeek 2017

[91] Basel Asmar, Andrew Day, Pritesh Patel, How viable are upstream projects?, CERAWeek 2018

[92] Basel Asmar, Andrew Day, Pritesh Patel, Cost & Innovation: Changing upstream dynamics, CERAWeek 2019

[93] https://www.hartenergy.com/exclusives/ep-executive-editor-bent-not-broken-187264

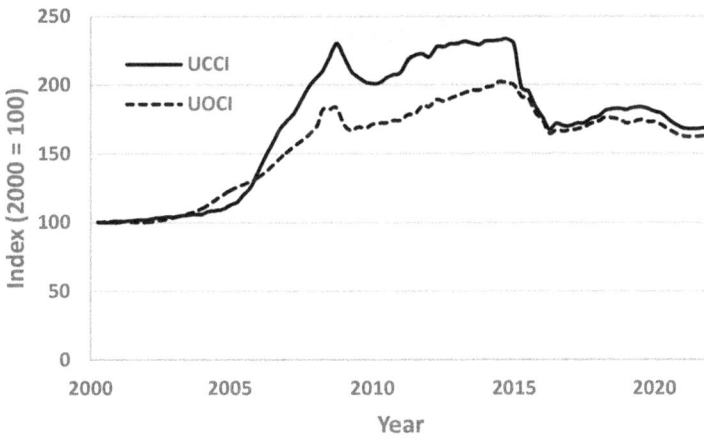

*Figure 3.16a: Upstream Capital and Operating Cost Indices*
Source: IHS Markit

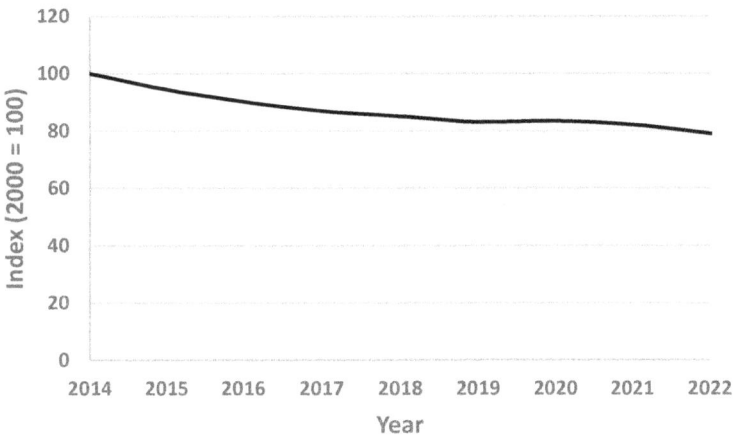

*Figure 3.16b: Upstream Innovation Index*
Source: IHS Markit

## Box 3: Cyclical versus Structural Costs

Oil and natural gas projects' costs are divided into two broad categories - cyclical and structural - based on their reliance on market conditions.

Cyclical costs are defined as the costs that are sensitive to market conditions and inflation and thus can fluctuate rapidly. These costs include costs of units (e.g., equipment, materials) and labour rates and are incurred as both capital expenditure (capex) and operating expenditure (opex).

Structural cost changes are cost changes that occur because of other factors. These cost reductions result from efficiency gains and design changes. The changes need not to be all technology-induced, but other practices such as simplification, standardisation, replication/repetition of processes and practices all play vital roles. Technological advances, including increased use of automation, digitalization, and widespread drone utilization, are implicitly included in and influence the contribution of both efficiency gains and design changes.

## 3.6   Energy Transition is Accelerating

The last decade has seen a rapid growth in focus on the "Energy Transition" by the world energy industry. Googling the term returns almost half a billion results as it has become more than a "buzz word". It is a major theme and agenda item in any energy or capital markets related discussion. However, "energy transition" means different things to different people, based on the sector, context, or location and particularly to the discussion is happening in developed or developing countries.

Nowadays, the term is increasingly used across various aspects of daily life. It is defined slightly differently in each sphere; however, its essential meaning is mostly the same. "Energy Transition" has become synonymous with policy actions associated with the fight against climate change. In very simplistic terms, it refers to the shift, by the global energy sector, from fossil fuel-based systems of energy production and consumption - including oil, natural gas and coal - to alternative systems. These alternative systems include renewable energy sources, such as wind and solar, as well as energy storage systems – mainly batteries such as lithium-ion batteries.

In the last few years, the momentum towards energy transition has been accelerating. The urgency to act has been reinforced by the climate change crisis and the increasing incidence of extreme weather experienced globally. Moreover, despite some loud vocal deniers of climate change (for example, many appointees within Trump's administration), the acceptance of climate change, as an emergency facing planet Earth, is becoming almost universal. with deniers getting moved into the fringes and being seen as pariahs.

Scientifically, it is now accepted that emissions, especially carbon dioxide and methane and other greenhouse gases (GHGs),[94] to the atmosphere are the main contributors to global warming. There is also broad acceptance that decarbonisation is essential to control the climate and reverse some of the damage that has already been done.

In 2015, many governments signed up to the Paris Agreement (or Paris Climate Accord), committing to reducing their countries' GHG emissions. This entered into force in 2016. This agreement within the United Nations Framework Convention on Climate Change (UNFCCC), commits 195 countries, and other signatories, to the goal of reducing carbon pollution, monitoring and reporting their fossil fuel emissions, with the goal of capping global warming to at least 2°C (3.6°F), ideally to 1.5°C (2.7°F). Many steps are underway to implement this, with many countries committing to time-framed, net zero emission plans.

While there is no doubt that transition is happening, the magnitude and speed at which this is happening is the subject of debate. The Covid-19 pandemic and its consequences point out that energy transition may have reached an inflection point, where visible (not public relations) responses to achieve it became mainstream, forcing governments and companies to take

---

[94] The main gases emitted, are greenhouse gases (GHGs), which are carbon dioxide ($CO_2$), methane ($CH_4$), nitrous oxide ($N_2O$) fluorinated gases (which include hydrofluorocarbons (HFCs), perfluorocarbons (PFCs)), and sulphur hexafluoride ($SF_6$).

it seriously or risk losing their reputation, customer base and power. There is intense pressure on politicians and business leaders to endorse a greener agenda, reducing the use of fossil fuels and divesting from oil and natural gas companies.[95, 96, 97] This pressure is being felt by oil and natural gas companies (even NOCs), with many shareholders (both individuals and institutional investors), and groups exerting pressure to transform their business practices. They are being asked to adopt Environmental, Social and Governance (ESG) criteria to decide future investments and even current business practices. Public opinion is turning against the traditional oil and natural gas business model. ESG is becoming the new focus, surpassing health and safety in forging new directions for companies. This pressure is driving energy transition and positioning it at the forefront of future development projects for the planet.

While energy transition is often positively portrayed in the media, we need to apply discernment before embracing it exclusively. In many cases, it can be a misconception, since not all acts of energy transition or alternative energy types are sustainable (See Box 4). As mentioned earlier in this section, energy transition relates to substituting the usage of fossil fuels with other types of fuels. Those alternatives are allegedly more sustainable. I say "allegedly" since sustainability of a certain energy type is debatable, depending on the point of reference in calculation. For example, when examining wind power sustainability, do we start from the mining process of the raw materials, the manufacturing, etc? Some solutions may appear good with green credentials, but when all the supply chain are considered, including mining and transporting materials, manufacturing, plus the excess power needed for all these tasks, these green credentials diminish considerably. With that in mind, renewables projects must consider their community impacts, i.e., any negative impacts on other communities that may suffer from mining for example, to

---

[95] https://en.wikipedia.org/wiki/Fossil_fuel_divestment

[96] https://oilprice.com/Energy/Energy-General/Goldman-Clean-Energy-Investment-Has-Hit-A-Critical-Point.html

[97] https://www.reuters.com/article/us-vatican-environment/vatican-urges-catholics-to-drop-investments-in-fossil-fuels-arms-idUSKBN23P1HI

evaluate overall sustainability. Stakeholders need to obtain legitimacy, credibility and trust before embarking on any project that will impact the lives of many people.[98]

---

**Box 4: Alternative Energy Classification**

"Alternative energy" is an umbrella term that encompasses the forms of energy intended to replace dominant fuel sources with fuels that have less impact on climate change and the environment. The exact definition of "alternative energy" has changed through time, with the dominant fuel changing from traditional biomass a few centuries ago, to fossil fuels (coal, then oil) now. Currently the term's definition differs between various data sources. While all agree that the dominant fuel type is fossil fuels, there is disagreement on which other types are excluded from alternative energy terms, with some specifically excluding nuclear energy or hydro-energy.[99] Throughout this text, the term "alternative energy" includes all forms of energy, excluding the three types of fossil fuels: oil, natural gas and coal.[100]

Thus, the following types of energy are considered forms of alternative energy:

- Nuclear energy, encompassing both nuclear fission and nuclear fusion.
- Hydro-energy, including traditional waterwheels, hydroelectric dams, tidal energy, wave energy and water current energy.
- Solar energy, harnessed for generating electricity or conversion into thermal, chemical or even mechanical energy.
- Wind energy, utilised to generate electricity or is

---

[98] https://www.petroleum-economist.com/articles/low-carbon-energy/energy-transition/2020/renewables-projects-must-consider-community-impact
[99] While I highlighted earlier my philosophical agreement with this. The term is used in this book to preserve consistency with the previous book, and to utilise the widely used terminology in literature.
[100] Both conventional and unconventional fossil fuel resources.

converted to mechanical power using windmills.
- Geothermal energy, utilised either to generate electricity or for heating (including heat pumps).
- Bio-energy, comprising solid biomass, including wood, liquid biofuels and biogases.
- Chemical energy, derived from chemical reactions, chemical solutions or retrieved from the difference in concentration between two fluids (e.g. osmotic energy).

The main Alternative Energy classifications are based on their origin, renewability and environmentally friendly credentials.

In terms of *origin*, all types of energy are divided into three categories: solar, terrestrial and lunar, derived by tracing the origin of all energy sources.[101] In terms of significance, solar origin contributes the overwhelming majority of energy sources, with terrestrial a distant second and lunar an insignificant third.

In terms of *renewability*, energy is either renewable or non-renewable. However, evaluating renewability is a grey area, as it is fluid and not well defined. This is interlinked with the third criterion based on *environmentally friendly credentials*, where energy types are divided into three categories: friendly, neutral, and unfriendly, i.e. ranging from harmless to harmful.

Two common misconceptions promoted by popular media are that alternative energies are always renewable and always environmentally friendly.

This is a fallacy, as several types of alternative energy are non-

---

[101] Attempting to place the alternative energy types into the three categories results in a classification that may surprise many readers and is briefly described below:
- Solar origin includes solar energy, wind energy, the majority of hydro energy (e.g. wave energy, current energy, part of tidal energy) and bio-energy as a result of photosynthesis. Note that fossil fuels also fall into this category, as according to the biogenic theory, they form due to the decay of plants and animals.
- Terrestrial origin includes geothermal energy, nuclear energy, part of hydro energy (due to gravitational energy) and chemical energy. Note that part of the natural gas falls into this category according to the abiogenic theory.
- Lunar origin includes part of hydro energy (majority of tidal energy).

renewable. Some are only renewable if their consumption is maintained at a sustainable level, where they are carefully managed, to avoid their depletion and maintain their consumption at a level allowing their replenishment.

Regarding environmental credentials, the claim that alternative energies are all friendly is, once again, far from true. Some types create carbon dioxide emissions, disrupt fauna and flora habitat, cause soil salination and land erosion. Many of these factors counter each other and, while one energy can be classified environmentally friendly based on one measure; it can be unfriendly based on the other.

For example, while it is almost universally agreed that fossil fuels are environmentally unfriendly, in terms of greenhouse gas emissions, there is still disagreement about whether various biofuels (perceived as environmentally friendly) emit more greenhouse gases, when the emission of their full life cycle is taken into account.[102,103]

Some, like wood have a significantly higher carbon density. In addition, deforestation resulting from commercial cultivation of biofuel producing crops, is causing additional carbon dioxide emissions. This negates the $CO_2$ absorbing qualities of the forests. It also causes competition between crops. i.e. either for energy or food, with many farmers opting to produce the former, at the expense of the latter. This choice is due to higher profitability and some claim it could lead to food shortages worldwide.[104] These "energy crops" are problematic as they are also competing for scarce water resources in many areas.

A second example is that of renewable hydroelectric energy, generated by dams. These structures can have significant negative impacts as they can accelerate soil erosion; increase soil salinity

---

[102] http://www.scientificamerican.com/article/industry-lashes-out-at-corn-biofuel-study/
[103] http://www.scientificamerican.com/article/controversy-over-biofuels-and-land-cut-from-ipcc-summary/
[104] Note that the use or arable land for energy crops is less a contributor to food shortages than the structure of the global agricultural system and an inadequate system of strategic grain reserves.

downstream in the rivers upon which they are constructed; interfere with the marine life in the rivers and estuaries; can cause high carbon dioxide emissions due the decay of algae and plants.[105] Similarly, wave and tidal turbines also interfere with marine life, while wind energy negatively affects birds' habitat.

It is apparent that all these alternatives have some environmental drawbacks.

Not only can energy transition be achieved by altering energy production, but also it can also be achieved by reducing energy consumption, thus de-stressing the demands on Earth's resources and reducing harmful emissions. Thus, to assess a certain path of energy production, it is essential to evaluate the type of alternative energy in terms of its sustainability. In addition, it may be prudent to consider a more feasible path, which is to stick with certain types of fossil fuels to produce energy. This can be achieved by applying mitigating actions (such as carbon capture and storage, or planting trees), so as to maintain reliable secure source of supply. In terms of consumption, we need to consider ways of reducing energy consumption by means of increased fuel efficiency, better insulation, reduced dependency on internal combustion engines in transport, etc.

Currently there is a massive amount of news coverage and publicity regarding the "cause" of alternative energy technologies. Supporting this cause is seen as responsible or "sexy". We rarely hear about the efforts made to curtail energy use. They are not well publicised since they are less fashionable, but they are more important than new alternatives in reducing the harmful effects of fossil fuels. An undeniable fact is that with increased human development, energy use will continue to climb, hence we need a concentrated effort to find efficient ways to reduce energy usage per capita, while maintaining comfortable lifestyle.

---

[105] http://www.climatecentral.org/news/hydropower-as-major-methane-emitter-18246

Admittedly in recent years, concentrated efforts were made to change the energy platform, to make it responsive to concerns about pollution, the environment and efficiency, but change was not happening quickly enough. 2020 changed all that, bringing energy transition to second place on people's agenda, driven by the number one item: the Covid-19 pandemic. The Covid-19 pandemic has accelerated energy transition and support for the Paris Agreement on climate change is solid. The Biden administration's decision to re-join the Paris Agreement, bringing the US back on board after the Trump fiasco, restored some optimism that policies could successfully be implemented to tackle the global climate crisis. However, while political intentions may be well meaning, unfortunately, the words speak louder than the actions. Politicians have not yet calculated how they can afford, or implement, many of the plans and objectives committed to in the Paris Agreement. These goals cannot realistically be achieved, without exerting huge economic effects on the populations of the world. In fact, the world's governments, financial and capital markets have yet to reconcile how the huge costs to mitigate the impact of climate change are to be funded, despite knowing that the cost of doing nothing will be higher than the cost of the remedy. Truthfully, ignoring oil versatility as an energy source is hard but the untold truth is, that to achieve the targets of the Paris Agreement, it is essential to develop technologies and materials that do not exist yet.

The fundamental question often asked is whether alternative energy sources can replace fossil fuels, while allowing humanity to continue to develop? With current technology, it is a resounding NO. Compensating for oil flexibility will be hard in the short and medium term. In the future, it can be achieved, but not yet. In order for energy transition to succeed, the trajectory has to be gradual. Starting with enhancing energy conservation, concentrating on both efficiency and demand side management and then optimising pathways to replace fossil fuel-based energy with renewables.

The truth is that for each sector that uses energy, the time scale will differ, and the practicality of the substitution will differ.

While the time scale may be only a few years in some sectors, e.g. electricity generation, it may be measured in decades in other sectors e.g. mining.

Let's consider electricity and power generation. In this sector, alternative energy sources made massive strides, and, in many countries, renewable energy is leading the way. The cost of alternative energy is falling, and that trend is not changing. This is a benefit of the economics of scale, even though it is slowing down. However, for some alternatives like solar and offshore wind, it is forecast that, through advances in technology, costs will further reduce. Although in general, for alternative energies, further technological breakthroughs (new material, digitalisation, etc.), will be needed to make that energy even cheaper. Hence many technical battles have been won, and many cost battles have been won also, with costs of renewable energy now amongst the cheapest solutions, in what is being dubbed as the solar revolution.[106] See Figure 3.17. Thus, the economic argument that fossil fuels will always make better returns than renewables can no longer be made.[107]

However, despite these technical and cost breakthroughs, still other important considerations remain - affordability, scalability, security of supply and reliability of energy sources. The resilience of distribution grids needs to be enhanced to allow for flexibility and eliminate single points of failure. Thus, for the time being, there is still a need to make fossil-fuel based baseload sources available, to ensure smooth power supply. Although rapid developments in storage technologies are happening, more efficient solutions are required. Only with higher efficiency may we be able to forge a way ahead, where renewable energy can fulfil most needs which would indeed confirm that peak fossil fuel demand has been reached.

---

[106] What we have seen in solar energy, is that technological innovation, development and creativity are applied to establish cheaper sources of renewable energy proving that successful, profitable and long term solutions can be found.
[107] http://admin.petroleum-economist.com/articles/low-carbon-energy/energy-transition/2020/germany-targets-global-green-hydrogen-leadership

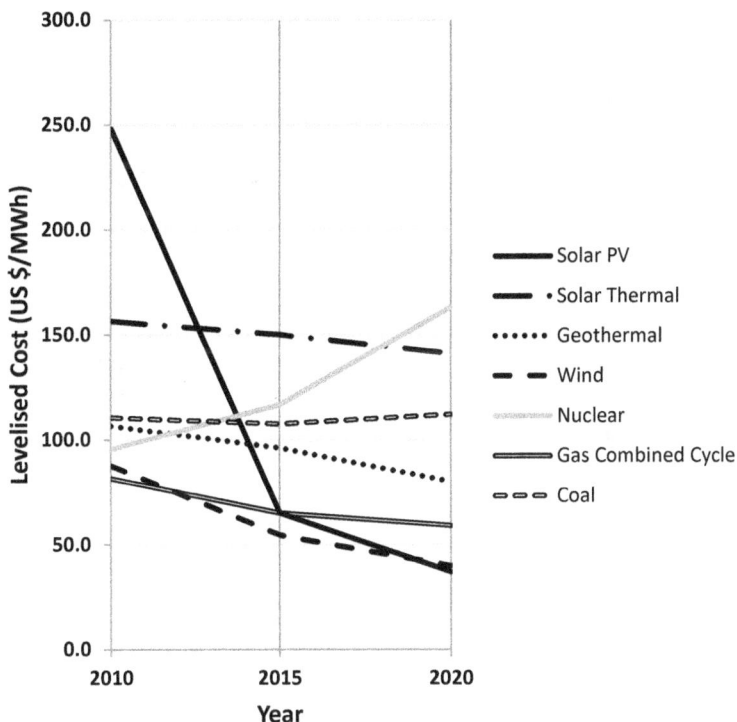

*Figure 3.17: Comparison of levelised or stabilized cost of energy analysis between 2010 and 2020 for selected energy sources*

Source: Lazard Levelized Cost of Energy Analysis, 2020, https://www.lazard.com/media/451419/lazards-levelized-cost-of-energy-version-140.pdf
Note 1: The levelized cost of electricity (LCOE), also known as Levelized Energy Cost (LEC), is the net present value of the unit-cost of electricity, over the lifetime of a generating asset. It is often taken as a proxy for the average price that the generating asset must receive in a market, to break even over its lifetime. It is a first-order, economic assessment of the cost competitiveness of an electricity-generating system, that incorporates all costs over its lifetime: initial investment, operations and maintenance, cost of fuel, cost of capital.

Now let's focus on the transport sector. Here the answer is more difficult assessing the suitability of alternative energy since, despite all the hype about electric vehicles, we still have several hurdles to overcome including battery storage, driving range, charging facilities, etc. Cracking the issue of transport will be the holy grail that will allow proper energy transition. I discuss this in detail in Chapter 5.

When we ask whether fossil fuels can be substituted, we need to consider the share of energy usage of fossil fuels in the overall energy demand. We need to assess if it is technically and economically feasible to replace that share with alternatives. The numbers tell us the facts, loud and clear.

Figure3.18 shows the world energy consumption from 1980 to 2019, where the upward trend shows no signs of reversing. 2020 was an anomaly, since due to Covid-19 pandemic, global consumption dropped. This is temporary and the upward trend is expected to continue.

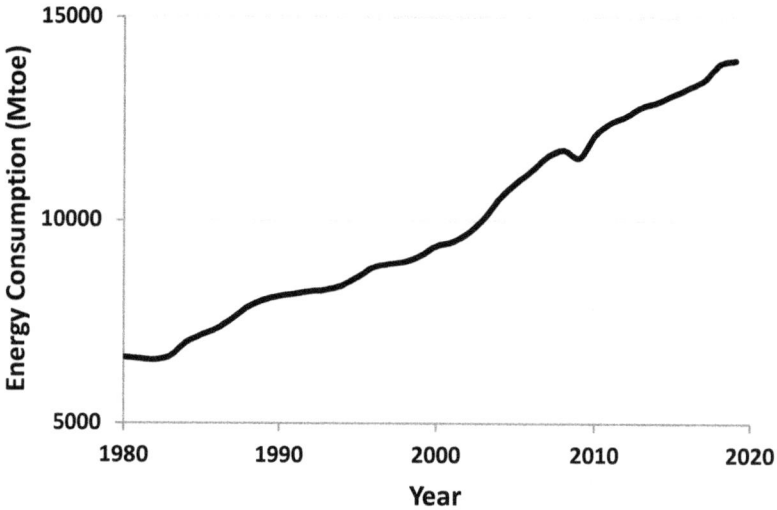

*Figure 3.18: World energy consumption (1980 to 2020)*
Source: BP Outlook 2016; BP Statistical Review 2020

In 2010, fossil fuels accounted for 87% of total energy demand. Between 2010 - 2015, the share of alternative energy in the global energy mix has increased by 0.9% to 14%, whereas fossil fuels accounted for 86%, so they were still dominant. By 2019 things improved slightly for alternatives exceeding 15%, but fossil fuels were still dominant, accounting for 85%. See Figure 3.19. The data shows that the share of fossil fuels in energy

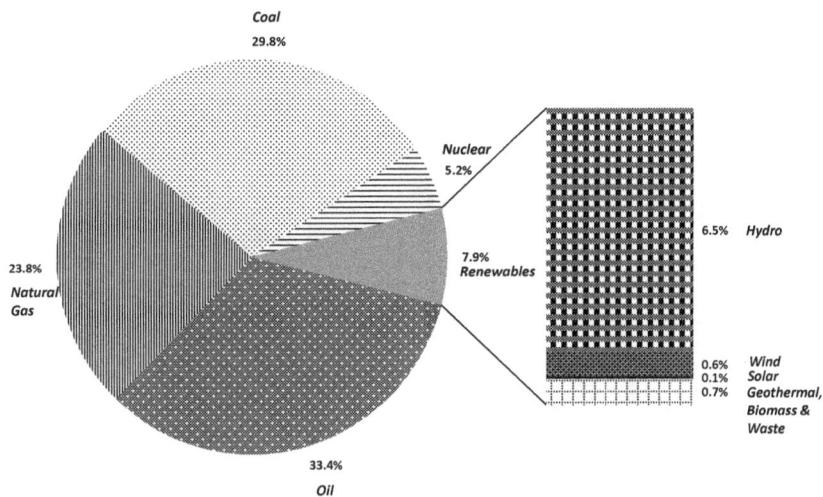

*(a) 2010*

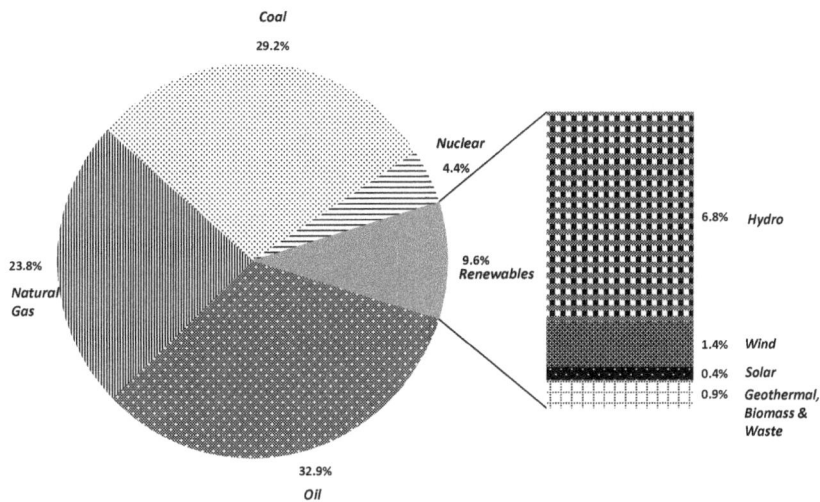

*(b) 2015*

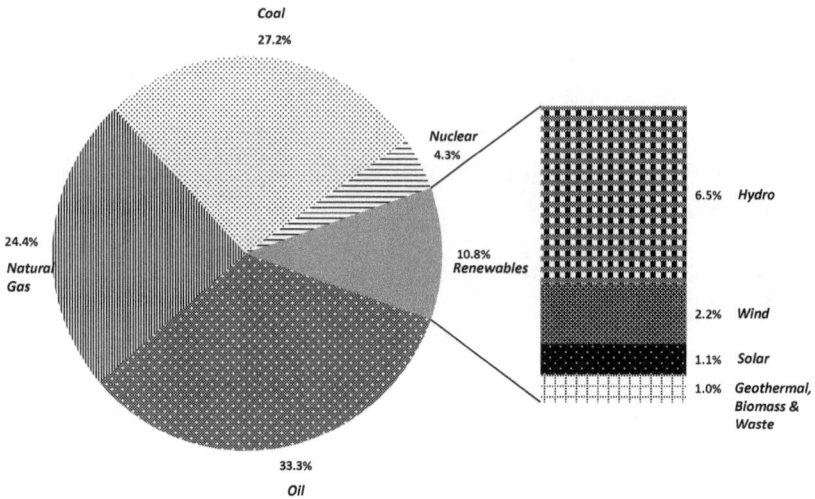

(c) 2019

Figure 3.19: *World Energy Mix (2010, 2015 and 2019)*
Source: BP statistical Review – versions 2010 to 2020

demand is declining. However, the decline is slow and if this rate is not accelerated significantly, it will take decades to eliminate fossil fuels.

It is clear that the share of alternative energy will increase in the energy market over time. The examination of future energy supply and demand scenarios authored by governments, think tanks, organisations and companies confirms this. The Paris Agreement sets out the roadmap to achieve this increase in the use, or development, of alternative energy sources.

The Paris Agreement roadmap has drawn from a variety of possibilities, offered by a host of organisations, governments and companies and most their opinions differ significantly. I am not going to discuss their merits or assumptions here, as this is outside the scope of this book. Interested readers can consult many published articles, e.g. scenarios from the IEA,[108] EIA,[109] OPEC,[110] BP,[111] Total,[112] IHS Markit,[113] Mckinsey & Company,[114] etc. There is a consensus - that until at least the mid of the 21st century, fossil fuels will continue to be the dominant supplying energy source but disagreement on the percentage of the share; hence the degree of dominance will be different, but overall renewables will account for less than 40% in most scenarios. Almost all scenarios agree that oil and coal will decline. The natural gas role will be debatable, and many scenarios suggest it may still grow in the short and medium term, but its involvement will certainly be longer.

In a more likely scenario, natural gas may be used as a rapid solution, to cut carbon dioxide emissions, by replacing other fossil fuels (e.g. coal) in baseload power generation. It may also support the greater rollout of renewables by providing backup capacity to overcome intermittency and reliability challenges. Natural gas can be pivotal in not only correcting imbalances of renewables' intermittent output, but also to support their rise. Batteries can also play an important role in counter-balancing the fluctuations in output from renewable energies. However, in their current forms, they cannot provide an adequate backup solution. While batteries can store surplus energy from renewables, the storage is only suitable to balance short-term (measured in hours) fluctuations in supply or demand of renewable energy. In order to balance days, or months, of supply and demand, a more reliable baseload solution is needed. That is where natural gas is positioned at the moment and combining it with carbon capture utilisation or storage initiatives, could effectively turn it into a "green fuel", although this label is disputed (see Box 5).

There is consensus that energy transition is happening and

---

[108] https://www.iea.org/reports/world-energy-model
[109] https://www.eia.gov/outlooks/aeo/
[110] https://www.opec.org/opec_web/en/publications/340.htm
[111] https://www.bp.com/en/global/corporate/energy-economics/energy-outlook.html
[112] https://totalenergies.com/system/files/documents/2021-09/2021_TotalEnergies_Energy_Outlook.pdf
[113] https://ihsmarkit.com/products/energy-climate-scenarios.html
[114] https://www.mckinsey.com/industries/oil-and-gas/our-insights/energy-2050-insights-from-the-ground-up

whoever disagrees with this is in denial. The truth is clear as is said in this Arabic proverb "الشمس لا تغطى بغربال", which means the sun cannot be covered with a sieve. So, the question is not if the transition is happening, but rather with its timing - when it will happen, its pace who will win and who will lose.

In the last few years, we witnessed a change in the attitude of oil and natural gas companies, with many shifting to endorse a green agenda. Companies are seeking ways to balance cost efficiencies with green measures, while maintaining their social license to operate.[115] They are setting emissions targets, clear climate policies and net zero ambitions.[116] Many diverted some of their investments into renewable energy, moving capex from upstream, with many CEOs claiming this change will lead to creating new jobs and opportunities. However, there was marked difference in attitude towards the transformation between US and Europe,[117] and between IOCs and NOCs. Although, in many instances, NOCs will have to adhere to the policies of IOCs they partner with, as they are often dependent on financial support from international financial institutions, who require certain ESG credentials from their creditors. NOCS, like the IOCs, can also be exposed to hostile public relations campaigns if they are not viewed as climate-friendly operators.

The trend started with European companies led by Equinor,[118] BP,[119] Shell,[120] Total,[121,122] Repsol[123] and Eni,[124,125] who all made acquisitions of companies in renewable energy, utilities and

---

[115] https://www.hartenergy.com/exclusives/what-oil-price-crash-means-decarbonization-187423
[116] https://www.hartenergy.com/exclusives/oil-ceos-pen-open-letter-pushing-climate-change-action-187726
[117] https://www.worldoil.com//news/2020/5/12/supermajors-all-have-ambitious-and-widely-varying-net-zero-goals
[118] https://www.equinor.com/en/news/20201102-emissions.html
[119] https://www.bp.com/en/global/corporate/news-and-insights/press-releases/bernard-looney-announces-new-ambition-for-bp.html
[120] https://www.shell.com/energy-and-innovation/the-energy-future/shells-ambition-to-be-a-net-zero-emissions-energy-business.html
[121] https://www.total.com/media/news/total-adopts-new-climate-ambition-get-net-zero-2050
[122] https://www.worldoil.com//news/2020/5/5/total-pledges-to-be-carbon-neutral-by-2050
[123] https://www.repsol.com/en/press-room/press-releases/2019/repsol-will-be-a-net-zero-emissions-company-by-2050.cshtml
[124] https://www.eni.com/en-IT/low-carbon/strategy-climate-change.html
[125] https://www.edie.net/news/6/Oil-giant-Eni-targets-net-zero-carbon-emissions-by-2030/

distribution networks. US companies did not show the same appetite initially and, despite showing early signs of action, are still way behind, remaining slow to move in this direction, with Occidental and ConocoPhillips leading the way.[126] However, their momentum in this regard accelerated recently, with other large players in the US and worldwide, including ExxonMobil, and Saudi Aramco, joining the club.[127]

Although many portray the actions as good deeds and justify them to demonstrate their moral responsibilities, the reality is many companies were forced to do this (due to public and investors pressure, regulatory obligations, and business necessities and realities). The shift toward cleaner forms of energy is forcing oil and natural gas explorers to cut costs and cancel, or defer plans, as uncertainties mount. In addition, digital transformation is providing a boost to energy transition, by allowing companies to reduce their carbon footprint in more efficient, cost-effective ways[128] and the new decarbonised or greener "Energy Evolution" businesses are less capital intensive and less risky.[129]

In fact, the renewables sector was one of the few that shrugged off the devastating effects of Covid-19 and grew in 2020.[130] Its share of investment is forecasted to overtake upstream oil and natural gas in 2021.[131] Although there were fears that the pandemic may risk slowing the energy transition,[132] after initial slowdown and job cutting,[133] the sector proved resilient and rebounded,[134] In fact in October 2020, there was a symbolic

---

[126] https://insideclimatenews.org/news/12112020/two-us-oil-companies-join-their-european-counterparts-making-net-zero-pledges/

[127] https://www.theguardian.com/business/2020/jul/16/worlds-largest-oil-firm-joins-alliance-to-reduce-industrys-carbon-emissions

[128] https://eloqua.upstreamintel.com/LP=27640

[129] https://www.hydrocarbonprocessing.com/news/2020/06/eni-ceo-working-to-make-the-green-drive-irreversible

[130] https://www.ft.com/content/205e5a30-4aba-4b9e-b27d-cc8a654ff684

[131] https://oilprice.com/Energy/Energy-General/2021-The-Year-Of-Renewable-Energy.html

[132] http://admin.petroleum-economist.com/articles/low-carbon-energy/energy-transition/2020/pandemic-risks-slowing-the-energy-transition

[133] https://www.eenews.net/assets/2020/05/14/document_ew_03.pdf

[134] http://admin.petroleum-economist.com/articles/low-carbon-energy/renewables/2020/like-a-bat-out-of-hell-part-four-renewables-set-for-rebound

moment when Nextra (the world's largest solar and wind power generator) market valuation overtook ExxonMobil's market valuation.[135]

Recently, even though it has been around for years, hydrogen has been touted as the new saviour and THE energy of the future. Before we get too enthusiastic, an important fact to remember is that hydrogen is not an energy source, it is an energy carrier. That distinction is very important, as energy sources are primary energy forms that are found in the natural environment, which are either finite (e.g. fossil fuels), or renewable (or relatively renewable), including solar energy. On the other hand, energy carriers (sometimes called energy currencies or secondary energy) are energy forms which have been transformed from primary energy sources (e.g., gasoline, electricity, work and heat). Thus, as an energy carrier, hydrogen is derived from several primary energy sources. The energy industry uses a colour-coded terminology to identify its source. See Box 5.

Theoretically, hydrogen can solve three energy challenges[136] - it cuts industrial carbon emissions, enables emissions-free heating and can store renewable energy. However, prior to having hydrogen as a widely adopted solution, in practice we need to look at the economics involved, being pragmatic and not idealistic. While hydrogen will, no doubt, play a role in the energy transition and has a role in the energy mix in the medium term, it certainly is not messiah. It has serious limitations, e.g., its low energy density, along with the need for significant investment in new and adapted distribution infrastructures. The technology to enable hydrogen, as energy, to be scaled up exists already, but its adoption will depend on the policy frameworks developed by governments to implement their decarbonisation goals. At the moment, several projects are underway and there is massive momentum supporting new projects.[137] Cost trends and

[135] https://www.ft.com/content/39a70458-d4d1-4a6e-aca6-1d5670bade11

[136] https://oilandgas.mhi.com/stories/how-hydrogen-can-solve-these-3-energy-challenges/?utm_source=Newsletter&utm_medium=email&utm_campaign=OG_Newsletter013&utm_content=3-Energy-Challenges

[137] http://admin.petroleum-economist.com/articles/low-carbon-energy/energy-

capacity deployment trends suggest a future competitive role for the fuel.[138] Energy companies are betting on a hydrogen boom, but that time is still some distance away.

Other energy carriers such as ammonia and methanol are also being promoted as part of the hydrogen economy. Although they can be more practical as they have higher energy densities, they suffer limitations, similar to hydrogen, and their use as possible mainstream solutions is currently overshadowed by the hydrogen hype, which has been grabbing all the headlines.

## Box 5: Hydrogen Colour Coding

Although Hydrogen is a colourless gas, different colours are assigned to it depending on the production type.

Three main types are often mentioned:

- **Green** hydrogen is produced by electrolysis of water, using only electricity from renewable energies.

- **Grey** hydrogen is extracted from natural gas in a process called steam reformation and is currently the most widely used method.

- **Blue** hydrogen is grey, brown, or black hydrogen, but with the emissions generated are captured and stored underground via industrial carbon capture and storage (CSS).

Other colour codes are:

- **White** hydrogen as the naturally occurring gas.

---

transition/2020/european-hydrogen-projects-make-progress
[138] https://www.petroleum-economist.com/articles/low-carbon-energy/renewables/2020/green-hydrogen-can-be-cost-competitive

- **Black/Brown** hydrogen is extracted from coal or lignite in a gasification process, converting organic or fossil-based carbonaceous materials into carbon monoxide, hydrogen and carbon dioxide. The carbon monoxide then reacts with water to form carbon dioxide and more hydrogen, via a water-gas shift reaction. This is the oldest way of producing hydrogen. Similar techniques are used with biomass as the raw material

- **Turquoise** hydrogen is produced by methane pyrolysis (thermal splitting). Instead of carbon dioxide, solid carbon is produced.

- **Pink** hydrogen is obtained by electrolysis through nuclear energy.

- **Yellow** hydrogen can be either hydrogen obtained by electrolysis through solar power, or, more commonly, it indicates that the electricity used for the electrolysis comes from mixed sources based on availability (from renewables to fossil fuels).

Besides natural gas, another "bridge", or a cornerstone, of many roadmaps to energy transition is considered to be carbon capture and storage (CCS) – see Box 6. CCS has been considered for decades. Since my involvement with the industry, I witnessed many occasions where CCS was promoted as the 'magic bullet' to save the environment, but then enthusiasm faded away. However, post-Covid-19, the time for CCS appears to have finally arrived, with many IOCs now taking it seriously as a definite way forward. Nevertheless, despite all the promise, we still need to face reality that CCS is effectively a tax and not a revenue stream. It is analogous to decommissioning – a necessary evil – in the eyes of oil and natural gas executives. As a result, operators emitting $CO_2$ (not only oil and natural gas operators, but also cement manufacturers, steel manufacturers, power stations etc.) will opt for the cheapest solution when, and if, they are forced to adopt CCS into their operations. Then, the option chosen by most will probably be a standardised, simplified and off-the shelf, if made available. Still CCS will only take off if there is an incentive i.e. by imposing a high carbon pricing[139] to force its adoption. When CCS is referred to as CCUS, by some who claim they can make a business case for it, (where "U" stands for utilisation), the claims need to be taken with extreme caution, as they can only materialise in limited cases. Thus far, the "U" has been an aspirational gimmick, unless it is for enhanced oil recovery (EOR), which ironically means it is part of a project which results in generating more emissions, but this may change in the future.

CCS is positioning itself as part of the future. With several high-profile projects announced, mostly in Europe[140] – it will undoubtedly grow into a new industry, some future estimates value it at over a trillion dollars annually![141,142] However we need to always remember that it is easier to by-pass energy generation operations to new alternative energy sources than decarbonise existing ones. The pathway may be cheaper and less controversial has more public support in many countries. CCS is competing with other forms of decarbonisation, including offsetting carbon emissions (e.g. reforestation). But many are not developing CCS as they are trying to move away from fossil fuels altogether.

## Box 6: Carbon Capture and Storage

Carbon sequestration, or carbon capture and storage (CCS), is the solution being touted as a solution to reduce carbon dioxide emissions. While technically possible, in its current form, it is cumbersome, complicated and expensive. It requires innovation to bring it up to the next level, i.e. transforming it to a practical

---

[139] "carbon pricing" could in fact be via a penalty (tax) or incentive (requirement to meet certain emission standards) ... etc
[140] https://www.hartenergy.com/exclusives/catching-carbon-186628
[141] https://pemedianetwork.com/petroleum-economist/articles/gas-lng/2021/ccs-could-be-trillion-dollar-industry-baker-hughes-am-2021
[142] https://www.bloomberg.com/news/articles/2021-01-13/occidental-oxy-wants-to-go-green-to-produce-more-oil

solution.

The aim of CCS is to prevent the release of large quantities of $CO_2$ into the atmosphere. It is a potential means of mitigating the contribution of fossil fuel emissions to global warming and ocean acidification.

CCS is simply the process of capturing waste $CO_2$ from large point sources, such as fossil fuel production facilities, power plants or industrial facilities, transporting it to a storage site and depositing it where it will not enter the atmosphere. Normally the storage site is carefully selected to be safe and offers a permanent storage site, located in an underground geological formation or exhausted gas field.

In process engineering technical talk, once $CO_2$ is captured, it is compressed into a fluid, then it is transported via pipeline, truck, rail or ship to the storage site, where it is injected deep into a rock formation that traps it.

Most publications refer to CCS from power plants, even when they mean general CCS. There are three basic types of $CO_2$ capture from power plants: precombustion, post-combustion and oxyfuel combustion.[143] The technical differences basically refer to when the process of extracting $CO_2$ from the gas stream is performed. Similar engineering principles are applied to capture $CO_2$ from industrial facilities.

Recently, several publications advocated using the term Carbon Capture Utilisation and Storage. The argument is that this may promote CCS more effectively.[144,145] The technical difference is that $CO_2$ can be utilised in other processes prior to its eventual storage. Often referred to as CCUS, the U is controversial as it can still contribute to producing fossil fuels as in EOR.

---

[143] http://www.thechemicalengineer.com/~/media/Documents/TCE/Articles/2015/891/891ccs.pdf
[144] http://www.aiche.org/ccusnetwork/what-ccus
[145] http://www.slideshare.net/UKCCSRC/ccus-in-the-usa-activity-prospects-and-academic-research-presentation-by-alissa-park

Furthermore, several companies are pursuing direct air capture (DAC), where $CO_2$ is captured directly from air, rather than from effluent streams, where it is then transported and stored. The process has been proven technically, but its cost remains prohibitive.

Another alternative energy that is transforming itself to stay relevant is nuclear, which faces its own energy transition.[146] While the fuel may appear to be in terminal decline, it is increasingly being accepted as part of a future energy mix, due to its emission-free nature. Advancement in designs and the promotion of smaller reactors means that it remains a possibility and is gaining acceptance as a plausible reliable alternative.

We are seeing the signals of the beginning of the end of the fossil fuel era. It is likely to be a long and dramatic journey, but it seems inevitable. In a nutshell, the conclusion is "combining natural gas and renewables is the fastest route to decarbonization".[147] Fossil fuels will continue to be a dominant feature of energy use through to the middle of the century, continuing to supply our energy needs alongside the growth of renewables, like offshore wind and solar power.

Rather than be vilified, when it comes to discussions about energy transition, the oil and natural gas industry should be part of the conversation, as it can bring so much useful expertise to the table. Unfortunately, with a low oil price, the ability of the oil industry to invest in research and development will be limited, which may unintentionally hamper the development of technologies needed for clean energy transitions around the world.[148]

---

[146] http://admin.petroleum-economist.com/articles/low-carbon-energy/nuclear/2020/nuclear-faces-its-own-energy-transition
[147] https://oilandgas.mhi.com/stories/why-combining-gas-and-renewables-is-the-fastest-route-to-decarbonization/
[148] https://www.worldoil.com//news/2020/4/15/clean-energy-transition-plans-imperiled-by-oil-s-crash?id=1774035

As part of the energy transition, one has to understand that setting a high "green premium" can hinder the net zero emission targets and thus the premium needs to be reduced to a realistic price level. The world needs to accept the hard fact that the 2°C Paris Agreement goal, while morally compelling, is unfeasible, unrealistic and is practically beyond reach. It will bring humanity back to "prehistoric" living standards (ok exaggerated figure of speech) – actually unacceptable to many communities and nations. A perfect solution is improbable but, in this case, it is better to have a half solution than no solution. Thus, perhaps a compromise is a hybrid model where the solution is an affordable path that will bring humanity closer to the aspired goal. In my opinion, we need to accept that combining natural gas and renewables is the fastest route to decarbonization. Energy transition will require time, capital and, since most renewables are more capital intensive, they require support from companies with muscular balance sheets, such as oil and natural gas companies, to be part of the formula to succeed.

All the steps taken so far will not be enough to replace the use of fossil fuels in the medium term. Energy demand is projected to continue increasing and the relative cost of fossil fuels development, compared to alternative energy, will continue to be competitive and will favour fossil fuels.

The post-Covid-19 era will be characterised by strained economies, where countries will scramble to pay for the cost of the pandemic and climate policies, in some countries, may take a back seat as economic growth will take priority. Thus, the outlook for fossil fuels depends heavily on governmental policies and choices, i.e. if governments pick the most affordable or the cleanest fuel? It is highly likely that behavioural changes and the social impacts of the pandemic result in a return to the energy demand trajectory which existed pre-Covid-19. But once the new norm emerges, governments will have to decide whether to use fiscal incentives to accelerate the energy transition or simply to kickstart the economy at the lowest cost. This will determine if oil still has a future or not.

We need to reflect on the fact that the oil market has experienced a series of entirely new events in the space of just a few months: the market was flooded with oil as a result of the breakdown in talks between Russia and Saudi Arabia, then demand vanished as a result of Covid-19 pandemic. This was both unprecedented and unpredictable and we cannot rule out another unexpected event in the journey of energy transition.

While the energy transition trend is inevitable, one scenario that has to be considered is that another oil price collapse, in the short or medium term, may threaten the renewable shift by swinging the economic feasibility back in favour of oil and natural gas. Thus, a post-Covid-19 oil price collapse may increase oil consumption and punish renewables, which will lose the low-cost advantage.

In fact, continued volatility in oil price (for various reasons) can have a destabilizing effect on the pace of development of alternative energies, particularly given the conflict of interest where integrated energy companies finance transition technologies and yet their revenues currently depend on oil and natural gas.

Furthermore, a global recession can make it difficult to secure financing for planned renewable projects.[149] They may lose the cost advantage, or worse, losing investment incentives may bring about an existential question in the aftermath of the Covid-19 crisis, which is, *should we completely turn our backs on one emergency while we deal with another? Or should we see this as a chance to rebuild the world into something better?* In my view, the answer should be a compromise and a gradual solution. We need to adapt a balancing act between the effects of Covid-19 pandemic and the climate crisis, to tackle both short-term pressing issues and long-term consequences.

Einstein once said, "We cannot solve our problems with the same

---

[149] https://www.thechemicalengineer.com/features/a-balancing-act-covid-19-and-the-climate/

thinking we used when we created them." This is true for us as we face into the reality that energy transition will be really hard and will require many sacrifices that the general public, despite everything, is not yet ready to accept. Thus, while aspirations to decarbonise are legitimate, the pathways will still be governed by realism of the laws of physics.

I am loathed to be the voice of doom, but I have to mention the white elephant in the room – methane fugitive emissions and leaks, as well as permafrost thawing releasing huge amounts of methane, may make all this energy transition pathway futile!

Perhaps there is also the fear that we will make the wrong decision and that the actions we take will not be sufficient to avert global catastrophe? Doing nothing is not an option. Dr Mike Ryan, of the WHO, said of the global emergency response to the pandemic, "Speed trumps perfection". The same is true here, we have reached the point where there is no room for failure, since failure to act quickly means this more devastating, secondary effect of global warming will not be mitigated.

## 3.7    The Balance of Power is Shifting Away from Petrostates

For decades, the image of petrostates was associated with wealth and prosperity, but with the march towards energy transition and the volatility of oil price, this perception is being shattered and the narrative of a new energy world will mean a change for those countries. It will result in some prospering, while undoubtedly, others will be left behind.

In my previous books I talked about winners and losers from low oil price. Here I revisit the discussion, noting that the landscape has changed dramatically in the last five years. The focus of the discussion has shifted towards energy transition and the battle between fossil fuel versus renewables. The spotlight will be on the minerals needed for a sustainable, prosperous, alternative energy, future world and the countries that supply it. Consequently, we will inevitably see the emergence of the power of the electrostates and chipstates, who will dominate the production of minerals needed for battery manufacturing, or semiconductors, respectively, who replace those previously powerful petrostates. Supply chain to net zero is critical and fundamental – the players who dominate it will be different, altering our geopolitical system.

While the eventual trajectory of the transition appears to be clear, the speed at which the world will reach it, is uncertain. It will depend on many interwoven factors and complex interplay of geopolitics, economics, as well as public health risks e.g. future pandemics.

If we accept that the shift will be gradual, fossil fuels will be here for years to come and some petrostates will transform themselves to maintain a significant role for the transition era and beyond. Leading those role players will be natural gas producers, who will benefit from the position natural gas will have in the energy mix. In the last few years, natural gas has been promoted as a bridge fuel leading humanity transitioning its energy usage, into greener alternatives. Which means that, unlike oil, there is still a degree of acceptance and good will towards the natural gas, which gives its producers hope for a place at the table, in the short and medium terms.

The gradual shift away from the use of fossil fuels also means that the role of oil and natural gas, in both the transport sector and as a petrochemical feedstock, will endure. I discuss the transport in more detail in Chapter 5.

It is vital we understand that politics, not technology, will determine the pace of energy transition and the actions to tackle climate change. Legislation and regulations are integral part of the jigsaw and, without appropriate incentives and enforcement powers, the drive towards a greener world will be slow.

There are realities and there is vision, when talking about this subject. The vision is always bigger, but not always completely achievable. In the last few months, we heard plenty of claims concerning the hydrogen future. For example, Goldman Sachs predicts green hydrogen will become a US$12 trillion market by 2050.[150] We also read all aspirational plans to reduce $CO_2$ emissions that go beyond CCS, i.e. advances in sunlight-activated catalysts or photocatalysts have generated extensive research in sun-powered chemistry, where the aim is to convert waste $CO_2$ into a raw material and sunlight is used as an energy source for producing various chemicals.[151] Similarly, efforts are being pursued in the construction industry to produce lower-carbon cement.[152] These efforts remain aspirations as they are currently too expensive. Science and technology can potentially cooperate to achieve these goals, but feasibility is a considerable remaining hurdle. Also, the political will to incentivise such solutions remain hesitant.

Another possibility is nuclear fusion, with the ITER[153] reactor assembly underway.[154] It is the holy grail of clean energy, but unfortunately, for many reasons, has not yet been realised.

In order to understand the scale of the alternative energy needed to achieve a sustainable earth, I refer you to the Masterplanet vision, as published in Time magazine.[155] If we are to meet the total world energy demand by carbon-neutral resources, existing wind power installations would require 16.1 million $km^2$, solar - 5.6 million $km^2$ and nuclear - 109 thousand $km^2$ (increased to 9.1 million $km^2$, if we include safety zones around plants). These are vast figures. Theoretically, we would need to cover all of Russia with wind power, all the USA with nuclear power and all of the Indian subcontinent with solar power, to achieve the target of replacing fossil fuels. Similar studies also highlight the enormity of the task and illustrates how difficult it is to be achieved.[156]

---

[150] Scientific American, 323, no. 8, Dec 2020, p. 36
[151] Scientific American, 323, no. 8, Dec 2020, p. 30
[152] Scientific American, 323, no. 8, Dec 2020, p. 34-35
[153] ITER is an international nuclear fusion research and engineering megaproject, which will be the world's largest magnetic confinement plasma physics experiment.
[154] Scientific American, 323, no. 8, Dec 2020, p. 62-71
[155] Time Nov 2/Nov 9 2020, p. 76-81
[156] https://energycentral.com/c/ec/how-much-land-does-solar-wind-and-nuclear-energy-require

As discussed earlier, historically, oil prices experienced regular cycles of ups and downs. This has given the oil industry its boom-and-bust character, where companies and governments increase their spending to engage in big projects, when the price is high, but often appear to be caught unaware when the oil price drops. At the end of each bust cycle, we are told that the end of oil is upon us and that this time there will be no recovery. This doomsday vision, to date, has never materialised and the oil and natural gas industry re-emerges again, restructuring and becoming relevant again. This time, the momentum is with the green energy, but oil and natural gas companies can' survive by adapting and transforming themselves into green energy companies. They may even emerge as dominant players in the new energy landscape, with oil and gas companies utilising their vast amount of data, transforming themselves into technology companies in a world where data is seen as the new oil.

In previous price cycles (see Figure 3.3), identifying winners and losers from low oil price was relatively straightforward, where oil importing nations benefited while oil exporting nations lost. Several rules of thumb relating oil price to GDP existed.[157] A rule of thumb often quoted was "that the world GDP increases by 0.25% for each US$20 drop in oil price".[158] Lower energy prices are credited as the key driver. This rule of thumb is obsolete and is no longer considered valid. It used to be that oil price affected GDP for consumers directly and adversely. This link has weakened tremendously, with oil share declining in the energy mix and because of increased reliance on renewables, in some countries. Plus, increased oil production from the USA has turned the nation into an oil exporter, breaking its dependence on imported oil and thus the previous effects of oil price on its GDP. Other minerals and metals are now more significant, with increased value and importance in the equation.

The same can be said also about the importance of

---

[157] http://blogs.wsj.com/economics/2014/06/18/four-ways-to-gauge-how-oil-prices-hit-the-economy/
[158] http://blogs.lse.ac.uk/politicsandpolicy/falling-oil-prices-should-help-europes-ailing-economies-but-the-wider-implications-of-the-price-drop-remain-to-be-seen/

semiconductors and chips. The availability and production capacity of semiconductors exposed the dependency and vulnerability, in the future, to the new risk of bottleneck from monopolistic powers. Energy and mobility are both now technology businesses that cannot function without chips and, reduced supply of semiconductors, is becoming a national security and economic weakness.

It is remarkable to realize how cyclical and repeatable the oil boom/bust story is. Here I copy (in italic) some text from my previous book.[159] Please note that the narrative is still applicable now, despite the time that has passed. Except for the dates, the story is almost identical, (November 2014 is replaced with March 2020) while the rationale and outcome have altered because of the Covid-19 pandemic.

> *During the boom era till November 2014, the Arab world's major economies were the big winners. Their sovereign wealth funds amassed hundreds of billions in dollars, which were invested mainly on western stocks, infrastructure projects (often awarded to western companies), importing goods and amassing weaponry. However, the reversal of the price turned them to big losers on a current account basis, since the earlier gains were not banked but the majority were instead spent. Oil exporting countries including Saudi Arabia, UAE and Kuwait suffered a massive loss of revenues having to instigate spending cuts and reduce budgets. Besides scrapping or delaying infrastructure projects, several Gulf countries reduced fuel subsidies and introduced some taxation, e.g. value added tax, to increase their revenue streams.*

As a response to 2020 oil price collapse, the same actions were

---

[159] Basel Nashat Asmar, Fossil Fuels in the Arab World: Seasons Reversed – Oil and Politics Interplay in the Arab World, 2050 Consulting, London, UK, 2017

indeed repeated.

> *Saudi Arabia and its allies adopted a strategy of defending their market share in November 2014 anticipating Iran's reintroduction to the market. They insist that any cut in production has to be across the board and shared between all countries. To achieve this, they need not only OPEC support, but also support from Russia and the USA. Neither country will support, however. The former due to its unwillingness as discussed in the above paragraph, while the later due to the fact that its government is constitutionally banned from manipulating the market.*

Saudi Arabia repeated its mistake again in March 2020, flooding the oil market in an attempt to kill the shale/tight oil production in the US, to defend the Saudi market share. However, its plan backfired spectacularly and the oil price collapsed (see Section 3.2) as, it coincided with the Covid-19 pandemic and subsequent collapse in demand. Unlike in 2014, in this instance Saudi Arabia re-enlisted Russia into a renewed OPEC+ pact to cut production in order to stabilise the market. In addition, the US government was supportive and the Trump administration, ignoring its legal obligations (banning it from manipulating the market), shelved the NOPEC draft law[160] and openly cooperated in cutting production to stabilise oil market.[161] It is worth pointing out that the actual production cuts were done by the producers for economic reasons. However, it was notable that regulators in some states e.g. Texas, considered imposing compulsory production cuts.

In the previous book, the situation in 2015 showed that there was a draw between winners and losers as a result of the oil price

---

[160] https://www.aljazeera.net/news/politics/2020/4/28/%D8%AA%D8%B1%D8%A7%D9%85%D8%A8-%D8%A3%D9%88%D9%8A%D9%84%D8%A8%D8%B1%D8%A7%D9%8A%D8%B3-%D8%A3%D8%B3%D8%B9%D8%A7%D8%B1%D8%A7%D9%84%D9%86%D9%81%D8%B7

[161] https://www.hartenergy.com/news/continental-resources-halts-shale-output-seeks-cancel-sales-187224

collapse of 2014. In this cycle there were more losers than winners as despite low price depriving petrostates from their revenues consumers were locked down due to the Covid-19 pandemic and hence did not benefit from the low oil prices.

In conclusion, there will be a shift in power from petrostates to electrostates. There will be a move from liquid dependence on oil to solid dependence on minerals. The shift from energy security to minerals security is underway, but its pace is still uncertain, with the viability of aggressive clean technology being questioned.

With the expansion of energy transition there is a stronger more possibility that more oil and natural gas resources will stay in the ground and turn into stranded assets. Thus, we may have entered the last stage or sprint to the end where companies and countries with low costs may do the last round of investment to maximise their potential before it is too late. And some such as the Middle Eastern petrostates may look past the oil rout to become major players in renewables, especially solar power.[162]

---

[162] https://www.bloomberg.com/news/articles/2020-05-05/mideast-petro-states-look-past-oil-s-crash-to-chase-solar-power

# Chapter 4
# *ARAB OIL'S JOURNEY FROM CRUCIAL TO QUESTIONABLE*

In the previous chapter I looked at the main principles shaping the oil and natural gas world and govern those markets. I discussed the accelerating energy transition, the implications on the energy markets, identifying who may benefit from low oil prices and the shift in power from petrostates to electrostates.[1] My analysis showed that demand of fossil fuels will remain significant, albeit decreasing and that despite all projected increase in renewables, fossil fuels will continue to be a significant contributor in any new energy world paradigm. Several scenarios exist regarding the paths of energy transition, but none is certain. We, as individuals, businesses and nations, need to learn to tolerate that ambiguity in the short and medium term.

In this chapter I reach the crux of this book, focusing on the Arab world. As a result of my analysis, I have positioned the Arab states in the future energy geopolitical arena or "map", evaluating their contribution to a new energy ecosystem, with different energy sources and players. Taking a 360-degree view, I recommend ways that the Arab world can stay relevant in a new energy landscape and how it can preserve its position as a vital player in a world where oil is less popular.

I am using the same framework used in my previous book, that examined three main pillars, which in my opinion, constitute the cornerstones of a politically and environmentally stable world. These pillars are energy, defence and stability. Unless these three pillars are synchronised and working in harmony the societal

---

[1] Analogous to a petrostate, an eletrostate is a nation whose economy is heavily dependent on the generation, production and export of electricity, or electrical storage equipment (e.g. batteries).

structure will collapse. Venezuela is an obvious example.

Thus, to determine if the Arab world is at risk of becoming irrelevant in a new world energy order, I examine its contribution to the world's energy prime source budget analyse whether the stability of the region is still strategically paramount, i.e. to the degree that the rest of the world needs to engage in pacifying the escalating political and security imbalance in the there. Since it is hard to separate energy from security and defence, the analysis considers the effects of the combination of these factors.

The analysis identifies certain events which confirm that some signposts have already materialised (see Chapter 2). At the same time, significant choices emerged in other areas, which cautiously indicate optimism that there are some promising green shoots for potential significant change. Overall, to use a meteorological analogy, the forecast is not great. Conditions are obscured, cloudy, with tauz "طوز" (annual summertime dust storms, as known in the Persian Gulf) in the air, which suggests, in many aspects, that there is evidence that the Arab world may be in for stormy weather.

## 4.1    Causing or Easing Energy Market Turmoil?

As I said five years ago inn my last book:

> *"The turmoil in oil and gas markets in the last two years has once again confirmed the strategic importance of the Arab world as the most important oil and gas player. While discussions earlier showed that oil and gas market dynamics are changing rapidly, one important fact remains, i.e. that supplies from the Arab world's major oil exporters (i.e. Saudi Arabia, UAE, Kuwait, Qatar and Algeria) are reliable, and experienced no production disruptions. This is in contrast to production situation in many other oil suppliers, who at some point during the last five years had to cut or halt*

> *production (e.g. Libya, Iran, Iraq, Nigeria, and Venezuela)"*.

At the time of writing this new volume, I would suggest that the turmoil in oil and natural gas markets in the last few months has, once again, confirmed the strategic importance of the Arab world as the most important oil and gas player.

Saudi Arabia has once again proved that it has resilience, will and strength to cause havoc *and* influence the oil markets. Despite Trump's rhetoric about American energy dominance the Saudis proved they are still *the* major player in the oil markets and, even though their actions in March 2020 caused damage for all, those actions strengthened the country's position. Within a few months, their subsequent policies controlling oil production by leading OPEC+ alliance produced results. It is as if the initial policy was based on the Arabic proverb "علي و على أعدائي" which means, "I prefer to inflict damage on all, as I may still end up the winner at the end."

But despite short term triumph declared by Saudi Arabia, who declared that the era of "Drill baby drill!" is gone forever,[2] the main questions now are a) was that policy correct? and b) will it benefit Saudi Arabia and the Arab world in the medium and long term?

To answer these questions, in the following paragraphs, I will examine the factors that drove those decisions. Once again history has repeated itself. Despite all opinions to the contrary, Saudi Arabia has proved that it still has the ability to manipulate oil markets by flooding the oil markets whenever it wants. As presented earlier, when they did it before, the results were devastating to oil price, plunging it to unchartered lows in April 2020.

Previously, Saudi Arabia was an essential constituent of oil flow

---

[2] https://www.bloomberg.com/news/articles/2021-03-04/saudis-bet-drill-baby-drill-is-over-in-push-for-pricier-oil

control system, acting as the oil flow control valve, where it restricted excess oil flow into the market in order to stabilise it. However, in recent years it has twice acted to flood the oil market - aiming to lower oil prices in an aggressive manner; undercutting other producers, to gain market share[3] and attempting to remove tight/shale oil producers from the oil market. The first time this policy was applied, in 2014, it failed. Shale/tight oil producers, despite cutting production initially, rebounded once the oil price recovered and then expanded. Unfortunately, from a Saudi Arabian strategic point of view, shale/tight oil producers flourished rather than failed.

The second application of the policy in 2020 was a measured but limited success. The shale/tight oil industry did have to cut production, however, again, it did not fail. With oil price recovering, the shale/tight oil industry proved its resilience and started a slow recovery. However now, the days of easy access to capital have gone and fiscal restraint will force the industry to downgrade its own growth ambition. Saudi Arabia certainly succeeded in curtailing oil production in the US, where most now believe that the country reached its peak production early in 2020. But at the same time Saudi Arabia's actions failed to completely end the shale/tight oil industry. The industry survived with more discipline and is more robust, to lower oil price, than in the previous round.[4] It might have had to contract for a short time, but the source rocks do not go bankrupt. Geology is what it is and there are still some really good assets available.[5] On this occasion, Saudi Arabia had to cut its own production, coordinating with others in the OPEC+ framework to manage the oil market.

Saudi Arabia possesses the largest oil production spare capacity in OPEC and, in fact, the world. Previously it coordinated its

---

[3] https://www.worldoil.com//news/2020/5/6/undercutting-rivals-helps-saudis-grow-share-in-key-oil-markets

[4] https://www.worldoil.com/news/2020/5/5/oil-at-30-may-be-enough-to-revive-shale-activity-say-drillers

[5] https://www.hartenergy.com/exclusives/path-forward-kpmgs-regina-mayor-says-its-grim-shales-not-dead-187454

actions with two allies (Kuwait and UAE), who together, control the majority of OPEC's spare capacity (see Figure 4.1).[6] Thus, the trio could, in theory, increase their oil production to compensate for oil shortages from elsewhere. The trio has reduced as Kuwait is facing production problems, (due to difficulties in its oil production processes from excess water).

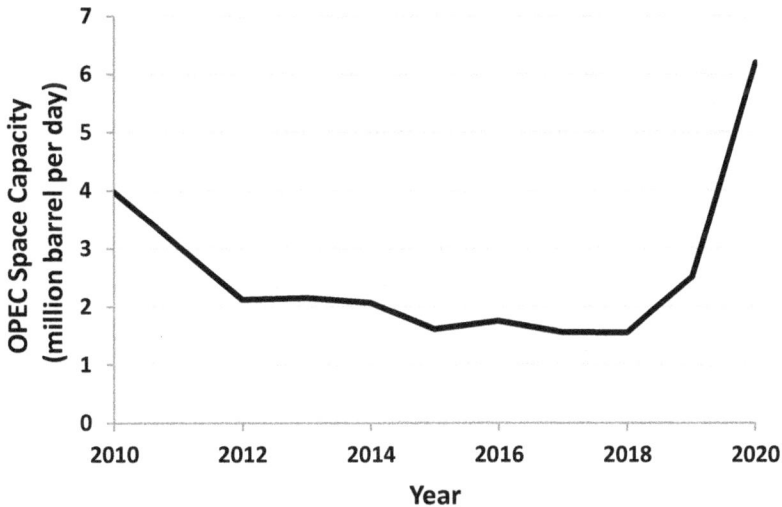

*Figure 4.1: OPEC average yearly spare production capacity (2010 – 2020)*
Source: EIA

However, since 2018, spare capacity ceased to be an issue as the debate shifted from peak oil supply, to peak oil demand and the pressing issue has shifted from lack of capacity to managing production cuts. As I mentioned earlier, this fact was illustrated when oil price reacted to a major Houthi attack on Saudi Arabia's oil production facilities in 2019.[7] Even when Saudi Arabia lost 50% of its production capability overnight, (following the claimed Yemeni attack on its main production facilities), oil price hardly moved, and a short-lived spike fizzled within days.

With so much oil supply destruction and the pandemic-caused

---

[6] https://www.eia.gov/forecasts/steo/report/global_oil.cfm
[7] https://www.nytimes.com/2019/09/14/world/middleeast/saudi-arabia-refineries-drone-attack.html

deferral of projects occurring in 2020 and 2021, it is inevitable that oil supply will be tight in the following years, putting upward pressure on oil price. This will be exacerbated by the "constraints to supply" imposed by OPEC+. Saudi Arabia and other petrostates naturally benefit, with additional revenues from higher prices. Also, the amount of restarted oil projects, following the forced shut-ins in 2020, due to both lower allocation of capital to maintaining supply and supply management, imposed by OPEC+, will be a vital oil price trigger, depending on the timing. Furthermore, there remains uncertainty about whether some projects will be shut down temporarily or terminated permanently.[8] Sceptics claim this may be the last opportunity to sell as much oil as possible, before demand collapses significantly. Time will tell. However, even if this is proven to be true, Saudi Arabia and its neighbours will have the last laugh, in the medium term, since their production costs are at the lower end of the cost curve.

Demand in natural gas crashed initially, following the Covid-19 pandemic,[9] but, as discussed in the previous chapter, natural gas is more resilient with projections of future demand being healthier than oil, Hence, major natural gas producing countries - including Qatar, Russia and the US, are pushing forward with their LNG expansion plans, while the rest of the world adopts a slower approach.[10]

Note however, natural gas markets remain regional and they are prone to seasonal price spikes. There is no equivalence of the OPEC+ framework to control natural gas markers. Many major natural gas markets are not part of OPEC (i.e. US, Australia). In addition in 2019, Qatar withdrew from OPEC (despite other minor oil producers joining), fearing a confrontation with Trump over the NOPEC law. Then voila! An oil price crisis occurred, and Trump facilitated OPEC+ negotiations to control oil markets and stabilise oil price.

---

[8] https://www.worldoil.com//news/2020/5/5/challenge-lies-in-choosing-which-wells-to-shut-in-amid-the-oil-downturn
[9] https://oilprice.com/reports/view/Global-Energy-Alert-12062019
[10] https://www.naturalgasworld.com/investing-through-the-downturn-lng-condensed-79394

It is often said that "numbers do not lie". Therefore, it is important to examine the numbers in the Arab world, in terms of oil and natural gas reserves, to examine if less dependence on Arab oil and natural gas is possible, or if the rest of the world is exaggerating the Arab world's importance.

As a result of the shale/tight oil and natural gas revolution and despite significant additional resources becoming economically feasible, it is the Arab world that continues to be a dominant force in oil and natural gas reserves, production and exports. (This is shown in Tables 4.1 and 4.2.) Additional historic graphs presenting the numbers in more detail, can be found in Appendix II.

*Table 4.1: Arab world oil and natural gas reserves share*

|  | Share (%) | | |
|---|---|---|---|
|  | **2010** | **2015** | **2020** |
| **Oil** | 50.4 | 43.4 | 43.1 |
| **Natural Gas** | 28.5 | 27.3 | 26.3 |

Source: EIA, calculated by the author

*Table 4.2: Arab world oil and natural gas production share*

|  | Share (%) | | |
|---|---|---|---|
|  | **2010** | **2015** | **2020** |
| **Oil** | 31.3 | 30.3 | 29.9 |
| **Natural Gas** | 15.4 | 15.6 | 14.2 |

Source: EIA, calculated by the author
Note 1: Natural gas production number till end of 2019.

The tables above demonstrate the significant share of reserves still located in the Arab world and illustrate that, while the region's status is diminishing, it remains a force to be reckoned with.

Yet, these numbers do not give the full picture. Not all reserves are equal, because they differ in terms of their quality and the cost to extract them. While it is hard to ascertain the share of the reserves to be left in the ground, it is reasonable to assume that the extraction cost will be a major determinant in the decision-making process of which resources to develop.

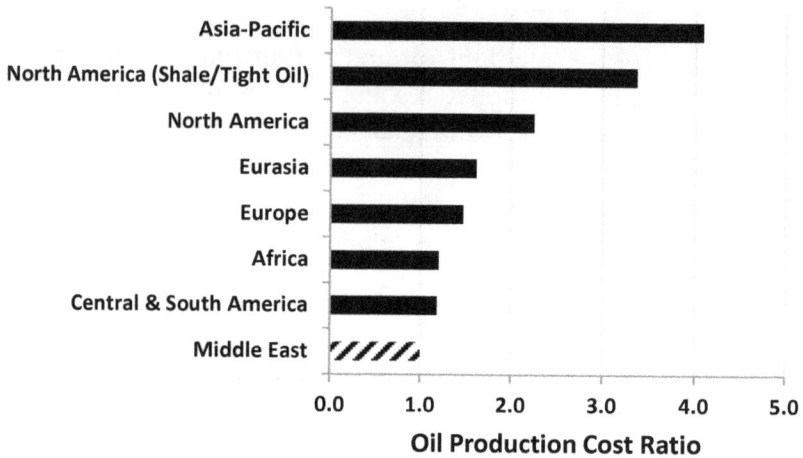

*(a) Onshore*

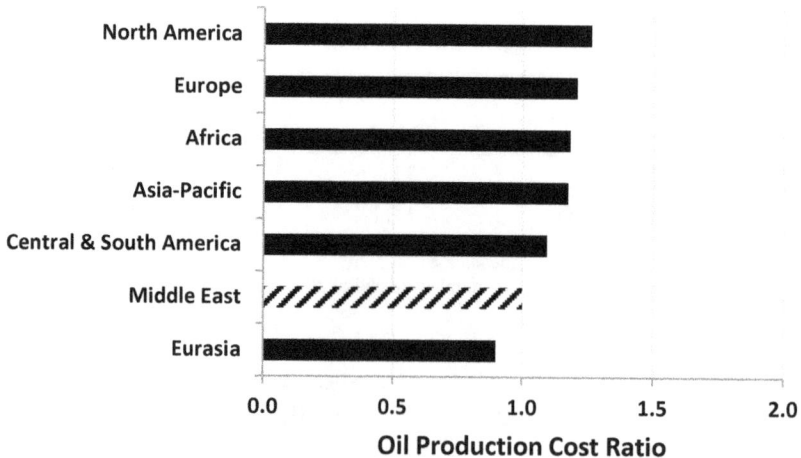

*(b) Offshore*

*Figure 4.2: Regional oil production cost relative to the Middle East (2020)*

Source: calculated by the author

Figures 4.2 through 4.4 show the relative costs for extracting oil and natural gas in the Middle East, relative to other regions in the

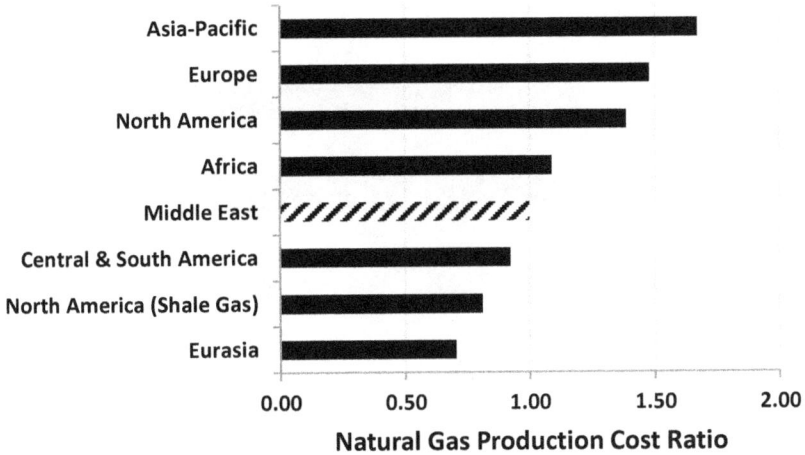

*(a) Onshore*

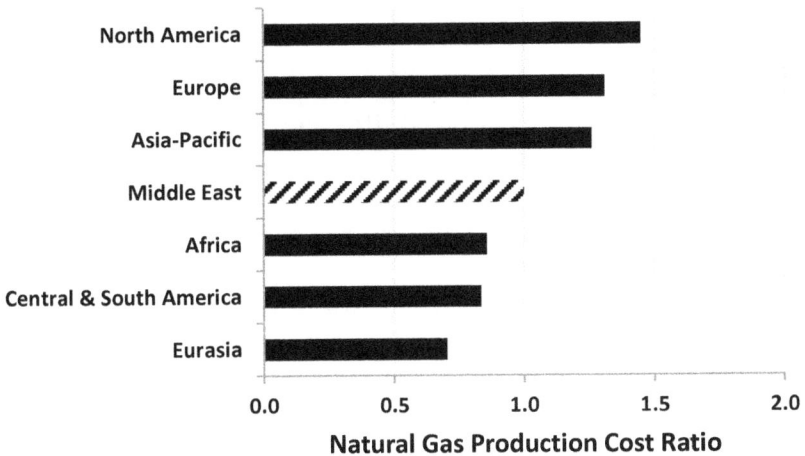

*(b) Offshore*

*Figure 4.3: Regional natural gas production cost, relative to the Middle East (2020)*

Source: calculated by the author

world. It is no secret that the region has lowest costs of oil production in the world.[11] However the graphs also demonstrate the magnitude of the margin.

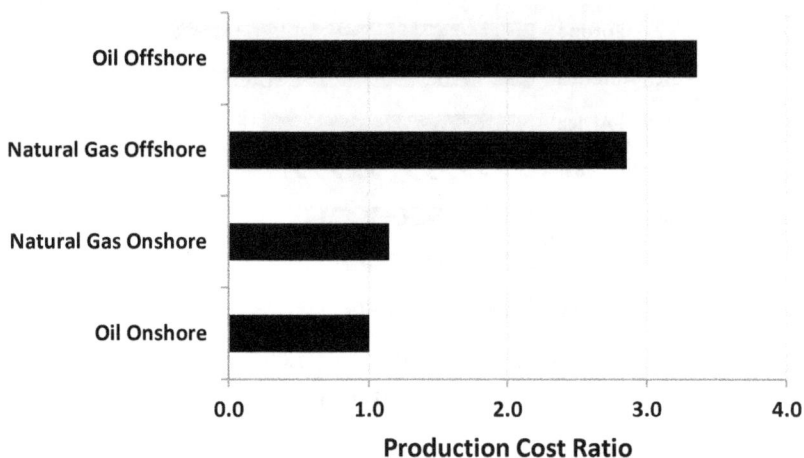

*Figure 4.4: Relative crude oil and natural gas production cost in the Middle East (2020)*

Source: calculated by the author

In order to allow fair comparison, the figures in the graphs were calculated based on undiscounted lifecycle development technical and engineering costs. They do not take into account any form of taxes, governmental licences, fees, environmental charges or other premiums. Nor do they include the cost of capital.

Note that this comparison differs from the many variations of breakeven cost curves often found in literature. These curves are becoming widely used and are often cited by oil and gas operators. Figure 4.5 shows a generalised oil production cost curve.[12,13,14,15,16] These types of "traditional" cost curves divide

[11] https://www.aramco.com/-/media/images/investors/saudi-aramco-prospectus-en.pdf, p.34

[12] https://www.rystadenergy.com/newsevents/news/press-releases/oil-production-costs-reach-new-lows-making-deepwater-one-of-the-cheapest-sources-of-novel-supply/

[13] https://askjaenergy.com/2016/02/04/current-low-oil-prices-are-not-sustainable/,

[14] http://www.edisonthoughts.com/2015/05/whats-marginal-cost-of-oil-supply-60bbl.html

[15] http://www.oilandgasmexico.com/2015/10/23/the-new-economics-of-oil/

the resource base by geographic regions, or by resource type (heavy oil, tight oil, etc). Other cost curves can be presented per country or terrain (onshore, deepwater, shallow water, etc.). While all these curves are informative, they can sometimes be misleading as they distort important data, e.g. concealing actual development costs and including exploration or acquisition costs The costs in such graphs are skewed heavily by all non-technical factors, including fiscal and legal factors.

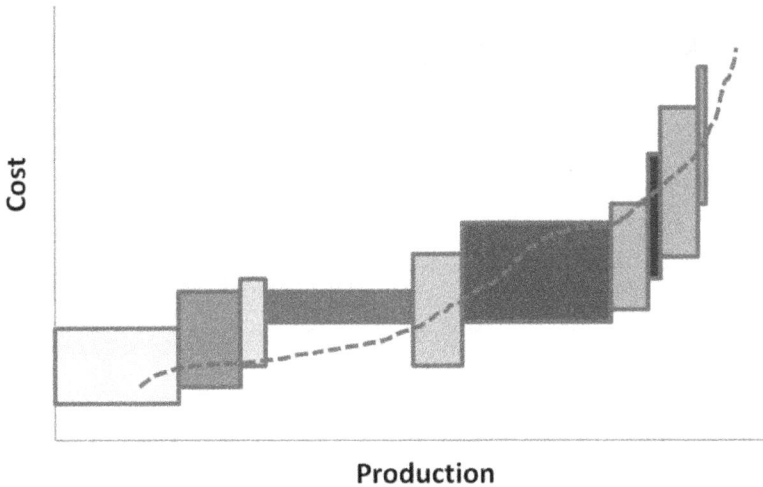

*Figure 4.5: Generalised oil production cost curve*

Note 1: illustrative sketch.

Thus, the comparison in Figures 4.2 through to 4.4, is more meaningful, as it compares actual engineering costs, which cannot be varied for political or other reasons.

It is significant that the collapse of oil price in 2020, pushed some unconventional tight/shale oil and natural gas outside the feasibility envelope/window. In technical terms, the reserves are pushed back into resources because the economics are not there. Hence, we need always to remember that reserves are elastic, with their number changing based on feasibility.

---

[16] https://www.woodmac.com/reports/upstream-oil-and-gas-global-oil-cost-curves-and-pre-fid-breakevens-updated-h2-2018-211878

From the graphs above we can see that the Middle East continues to have the lowest production costs in the world. Therefore, from petroleum economics point of view, Middle Eastern Arab countries can sustain production profitably, even at repeated and lasting low oil environment scenarios.

Consequently, the significance of Arab oil and natural gas reserves is amplified when the costs and the volumes of reserves are considered in tandem. The oil reserves will be the last to shut down and cease production if the price of oil remains low. This cost-volume combination is what allows the Arab world to punch above its weight and retain its' relevance. It also partially explains the continued interest in the region and gives the Arab world its realistic strategic value.

Although oil and natural gas exploration activities declined significantly since the Covid-19 pandemic, exploration activity was underway in many Arab countries prior to that. In fact, since 2015, several oil and gas discoveries have been announced in the Gulf countries, the Eastern Mediterranean and offshore Mauritania.

In the Eastern Mediterranean, significant natural gas discoveries were announced offshore of Egypt and Cyprus. The volumes discovered are expected to increase. Additional possible discoveries from the same basin are anticipated in offshore Lebanon, where several exploration licenses were awarded. Further potential exists in offshore Syria, where contracts have been awarded to Russian companies to start exploration activity. In addition, Turkey continues its exploration efforts in the region, having made a major natural gas discovery in 2020 in the Black Sea,[17] and it is considering campaigns offshore of Libya. Also, both Israel and Cyprus are active. However, as one would anticipate, territorial disputes are creating increased tensions in the region.

---

[17] https://www.atlanticcouncil.org/blogs/energysource/turkeys-gas-find-in-the-black-sea-how-big-is-this-tuna/

Another Arab area that is witnessing increased natural gas discoveries is offshore Mauritania which, alongside discoveries in Senegal, is being transformed into a new natural gas producing region.[18]

More significantly, the discoveries of massive reserves of natural gas offshore in East Africa - in Mozambique and Tanzania, have attracted interest by international oil and natural gas companies. The wider region is now considered a world-class natural gas play.[19] Oil and natural gas operators are acquiring licences for offshore the Comoros Islands and even offshore Somalia, since operators hope that natural gas resources will extend to Somali and Comorian regional waters.

In terms of both oil and natural gas, due to politics, the Arab world is hugely underexplored, with many countries were closed for international oil exploration. Major examples include Libya and Iraq.

In the Gulf region, significant discoveries were announced in Iran[20] and the UAE.[21] Although Saudi oil reserves growth has lagged the rest of the world, it nevertheless grew.[22]

In addition, recent estimates of unconventional oil and natural gas reserves in the Arab world indicate considerably higher volumes assessments than those I reported in 2015.[23,24,25] The estimates continue to increase, following announcements of major new discoveries, including new fields in UAE (22 billion barrels of

---

[18] https://www.oxfordenergy.org/publications/mauritania-senegal-an-emerging-new-african-gas-province-is-it-still-possible

[19] Gas play - An area in which hydrocarbon accumulations or prospects of a given type occur.

[20] https://www.dailysabah.com/opinion/op-ed/will-the-new-oil-discoveries-in-iran-change-energy-geopolitics

[21] https://www.worldoil.com/news/2020/2/6/uae-finds-80-tcf-gas-field-the-world-s-biggest-since-2005

[22] https://www.forbes.com/sites/rrapier/2019/02/22/saudi-oil-reserves-growth-has-lagged-the-rest-of-the-world/?sh=4478c7871da0

[23] http://www.eia.gov/todayinenergy/detail.cfm?id=14431

[24] https://www.eia.gov/analysis/studies/worldshalegas/

[25] https://www.worldenergy.org/wp-content/uploads/2016/02/Unconventional-gas-a-global-phenomenon-World-Energy-Resources_Full-report.pdf

resources)[26] and Bahrain (80 billion barrels of resources).[27] Note however, that many areas remain unassessed so these numbers will certainly continue to increase.

Despite these massive numbers, there has been only limited development of unconventional resources in the Arab world. The developments so far are all of natural gas reserves. These include tight gas developments in Algeria,[28] Egypt[29] and Oman,[30] as well as planned shale gas projects, plus ambitious development plans in Egypt[31,32] and Saudi Arabia.[33,34,35] No tight oil developments have yet been announced, despite significant potential, in Libya, Algeria and Saudi Arabia.

Regarding oil shale,[36] no efforts have been made yet to exploit these resources, except in Jordan[37,38] a country with limited conventional resources.

Following years of gaining very little leverage in the region, in the last few years, international oil companies gained more leverage from oil contracts in the Arab world. National governments still control the industry but are luring IOCs by

---

[26] https://www.gulf-insider.com/new-major-oil-discovery-in-abu-dhabi/

[27] https://jp.reuters.com/article/instant-article/idUSKCN1HB1ZJ

[28] http://www.hydrocarbons-technology.com/projects/timimoun-natural-gas-project/

[29] https://www.onepetro.org/conference-paper/SPE-94106-MS

[30] http://www.worldoil.com/news/2016/2/14/bp-extends-scope-of-oman-s-khazzan-gas-development-with-new-agreement

[31] http://www.egyptoil-gas.com/publications/the-future-of-unconventional-oil-and-gas-in-egypt/

[32] http://www.shalegas.international/2015/05/26/work-to-start-on-apache-shell-fracking-project-in-egypt/

[33] http://www.wsj.com/articles/saudi-aramco-close-to-awarding-500-million-shale-gas-contract-1456244375

[34] http://www.shalegas.international/2016/02/01/kingdom-of-shale-unconventional-developments-in-saudi-arabia/

[35] https://energy.economictimes.indiatimes.com/news/oil-and-gas/saudi-aramco-says-it-received-approval-for-jafurah-gas-field-development/74263532

[36] Oil shale is a misleading term (it is not shale and contains no oil!) that refers to organic-rich fine-grained sedimentary rock. It contains significant amounts of hydrocarbons in the form of *kerogen* (a complex mixture of hydrocarbon compounds of large molecules, containing hydrogen, carbon, oxygen, nitrogen, and sulphur), from which liquid hydrocarbons can be extracted through destructive distillation or exposure to heat into a form of crude oil termed *shale oil*, which subsequently may be refined into normal petroleum products. Oil shale differs from other oil deposits since it does not contain liquid hydrocarbons or petroleum as such, but organic matter derived mainly from aquatic organisms.

[37] http://www.shell.com/about-us/contact-us/contact-jordan.html

[38] http://blogs.platts.com/2014/04/04/jordan-shale-oil/

offering joint ventures and minority stakes in many new developments. Even in places like Iraq, which proved challenging to IOCs, operators are still exploring possibilities, despite a history of getting only meagre service fees for developing production facilities. In other countries, e.g. Libya, operators have written-off their investments.[39] However international service companies, engineering procurement and construction companies (EPCs) and equipment manufacturers have benefited. Nowadays, despite the spending cuts by global oil and gas companies, their percentage cuts in the Middle East were less than elsewhere. In many Middle Eastern oil producers, the upturn has already started, ahead of many other places, with major project announcements in Qatar and UAE.

In addition, in order to secure supply routes, Oman and Iran[40] are both planning, and engaged in, projects to export oil from the Arabian Sea and the Gulf of Oman.

However, lower oil revenues forced oil producing Arab countries, who had accumulated enormous sums of petro-dollars in sovereign wealth funds, to tighten their belts. They had to withdraw massive sums of money from these funds to be able to fulfil their cash obligations. This continued overspend, drawing rapidly on the future generations' security funds, is not sustainable and can lead to serious consequences.

In order to sustain the spending, previously unthinkable measures have been implemented. These include imposing taxes in the GCC and the floatation of Aramco, to create the world's largest listed company.[41]

Despite this great potential for growth in oil and gas supply from the Arab world, higher demand from within the region means that it is possible that exports will decline more quickly than

---

[39] http://oilpro.com/announcement/1549/libya-oil-write-off
[40] https://www.reuters.com/article/us-iran-oil-gulf/iran-plans-oil-exports-from-the-gulf-of-oman-to-secure-crude-flow-idUSKBN23W1M1
[41] https://www.nytimes.com/2019/12/06/business/energy-environment/saudi-aramco-ipo.html

anticipated. This effect will be counter-balanced by the fact that production costs from the Arab countries is the lowest globally and will continue to be so. Furthermore, with new policies cutting subsidies, demand will eventually slow, creating more export potential.

The double whammy of the supply glut and Covid-19 demand destruction, is a major threat to the Arab countries whose economies rely on oil revenue. The sudden decline in revenue exposed the structural weaknesses in the economies, due to the lack of diversification. The political leaders are now at a crossroads. They need to either be uncharacteristically transparent with their citizens, admitting that standards of living will deteriorate and infrastructure projects will suffer, or bury their heads in the sand and borrow, to buy themselves a few more years, knowing that the day of reckoning is only delayed or postponed and then face the consequences.

## 4.2 Dependence on Arab Oil and Natural Gas is Diminishing

In the last five years, oil trade dynamics have changed fundamentally, with the USA becoming a major oil and natural gas exporting country. The extensive development of shale/tight oil and natural gas plays in the USA, increased its oil production to 12.8 million b/d and 155.5 TCF/d in early 2020.[42] This has cut the USA's dependence on imported oil while the increase in natural gas production, transformed the country into a net natural gas exporter. In the same time frame, we saw a decreased coal production. Thus, this increased domestic supply in combined fossil fuel production, meant that the USA's net imports of fossil fuels (crude oil, natural gas and coal) decreased significantly between 2015 and 2020.

Besides the dramatic changes in the USA, the oil and natural gas trade picture has changed dramatically in the last five years. See Figure 4.6. The renewed American sanctions on Iran in 2018,

---

[42] https://www.eia.gov/petroleum/production/

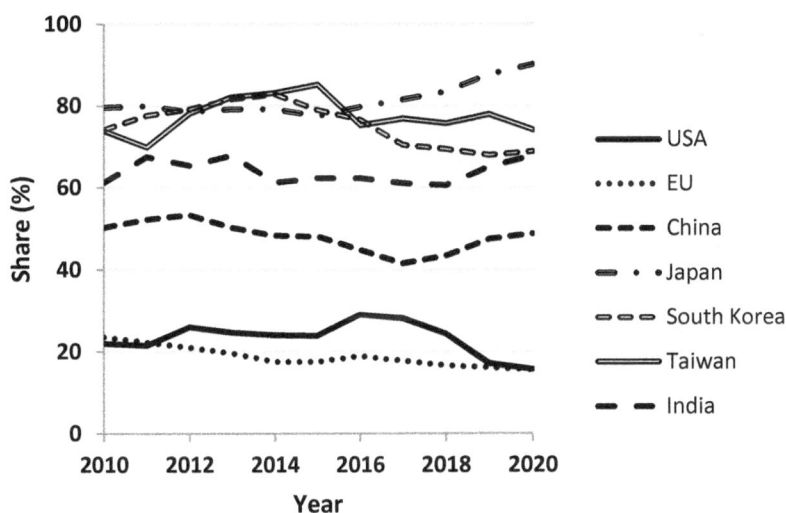

*(a) Crude oil imports from the Arab world*

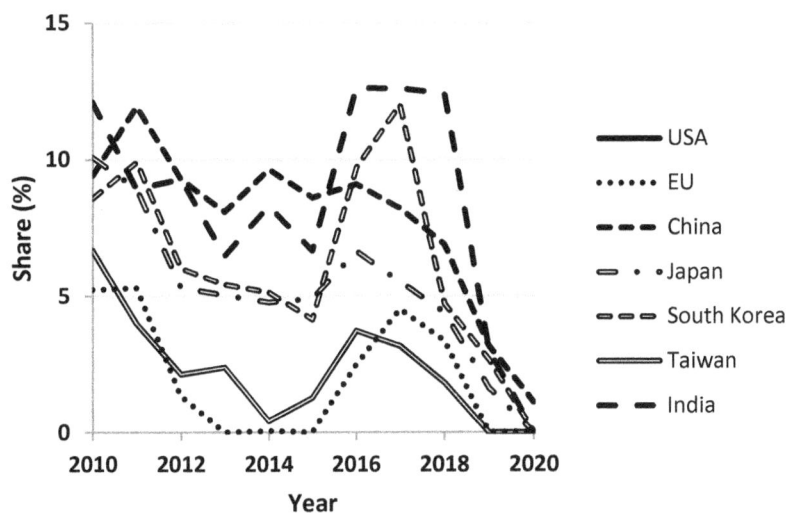

*(b) Crude oil imports from Iran*

*Figure 4.6: Crude oil and natural gas imports from the Arab world and Iran by the United States, EU, and several Asian countries*

Source: EIA; Eurostat; China Customs; Japan Ministry of Finance; Taiwan Directorate General of Customs; Korea Customs and Trade Development Institution; India Ministry of Commerce
Note 1: The United States had no crude oil imports from Iran.
Note 2: None of the covered countries had natural gas imports from Iran.
Note 3: Numbers are shown as percentage from the Arab world or Iran per importing country.

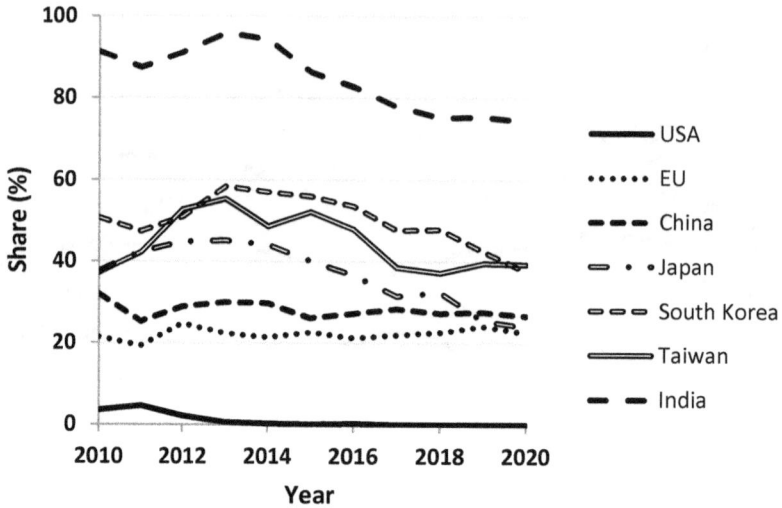

*(c) Natural gas imports from the Arab world*

*Figure 4.6 (cont'd): Crude oil and natural gas imports from the Arab world and Iran by the United States, EU, and several Asian countries*

Source: EIA; Eurostat; China Customs; Japan Ministry of Finance; Taiwan Directorate General of Customs; Korea Customs and Trade Development Institution; India Ministry of Commerce
Note 1: The United States had no crude oil imports from Iran.
Note 2: None of the covered countries had natural gas imports from Iran.
Note 3: Numbers are shown as percentage from the Arab world or Iran per importing country.

deprived the country of major export markets and its exports almost stopped completely, bar a handful of exempted or defiant countries.

In addition, unrest in Libya and Venezuela almost wiped them out of the oil market picture. Other minor producing countries, such as Yemen and Syria, continue to be out of the picture for the time being as well.

As can be seen in Figure 4.6, the USA's reliance on the Arab oil and natural gas all but disappeared, while its previous reliance on Iran is now a thing of the past.

Similarly, as seen in Figure 4.6, the EU does not rely on Arab or Middle Eastern oil and natural gas. The EU block has been flexible, diversifying its oil and natural gas resources, although many of its members still have an unhealthy reliance on Russia.

While the USA and Europe are less reliant on the Middle East for their energy needs, Asian countries (e.g. China, India, Japan) continue to rely heavily on the Middle East for their oil and natural gas supplies. Some are trying to diversify supplies to improve their security of supply, thus freeing them from depending fully on the Middle East in the future.

Note - the above paragraph paints a picture, prior to Covid-19 pandemic, which has had a significant, short-term, impact on the demand numbers. With demand destined to gradually recover, the dynamics of the trade are not expected to alter significantly in the medium term.

With the USA trying to disengage from the Middle East and concentrating instead on Asia-Pacific, several Asian countries are engaging with the Middle East, stepping in fill the gap. China has increased its efforts to get involved in the region's affairs and India is concerned about its dependence on the region and has made noises but has had made little progress in its efforts to diversify.

In the last few years, concentrated efforts were made by Asian countries to diversify their oil and natural gas supply options. For example, China signed several bilateral agreements with Russia to import natural gas. At the same time, it sponsored Central Asian countries and started importing natural gas from them. It also constructed numerous LNG import terminals to source natural gas globally - from Qatar, Australia, the USA, etc. In addition, it has committed itself to increasing oil imports from the USA.

Similarly, Japan is diversifying its LNG supply to include the USA, Australia and Russia as additional suppliers.

Five years ago, I wrote that we needed to keep an eye on American oil exports and natural gas exports, monitoring how they will come into the energy markets' equation. What we have seen is that the USA became a significant player in oil and natural gas export markets, with their exports increasing significantly. Obviously, the ambitious American expansion plans have been curtailed by Covid-19 pandemic, but we still need to watch the developments in North America. Experience has taught us that the players there are resilient and always will find ways to remain relevant. See Section 3.5.

Nowadays we hear scaremongering commentators, warning of impending supply gap, especially in oil markets. While I believe these warnings are exaggerated, I am confident that, if there were to be an oil supply gap, it will be filled by low-cost producing Middle Eastern countries in the short term, where spare capacity still exists. The reports about a shortage of capital expenditure (capex) for upstream and the resulting impact creating a shortage of oil and gas supply, fail to recognise that costs have come down considerably, enabling many projects to breakeven at lower oil prices and start operating with less capex.

Moreover, many Western companies have set different hurdle rates for projects based not only on project profitability, but their stated intent to decarbonize. Thus, some of the projects affected may be divested and instead, operated by others, but there is at least some slowdown of investment. For example, In North America many projects are highly profitable at current oil prices, but have not yet had the amount of additional spending one might expect. Unfortunately, profits are going to shareholders, not to generating new supply. This will ultimately create potentially serious consequences

In the medium term, there may be more reliance on the Arab world to fulfil global oil supply as other producing countries cannot compete in terms of costs. But with demand projected to peak and then decline, this dependence will not be as critical as it is these days, since it is hoped that the energy transition will result in an increased share of alternative energy sources filling the gap.

## 4.3 Renewable/Alternative Activities – Can the Arab World Supply Renewable Energy Successfully?

Initially when I started writing this section, I was pessimistic, believing firmly that the Arab world has certainly missed the boat and the end of oil age will plunge the region back into the dark ages. However, I was a little more optimistic when I discovered the activities being made in the renewable energy arena especially with ample supplies of solar energy, which reveals a potentially a bright future for the Arab world in the new era of energy transition, **if** the governments of the region play their cards right.

Although there are many renewable energy plans & projects in development, too many to list here, I believe success will depend on a significant change in mentality from those in charge. Historically, governments in the region have focused on public relations campaigns boasting of building the largest of their kind projects, to an entirely different mindset. Government leaders and decision makers need to create integrated, innovative, comprehensive plans if they wish to become major players in the new era.

While the renewables drive is gaining momentum in the Arab world, most efforts relate to the power generation and electricity sectors.

Furthermore, as part of the energy transition efforts, policies have been announced regarding the adoption of new global trends in the transport sector. However, all these policies depend on imported technology, with little technical or research and development efforts being sourced locally. The mobility and transport sector is discussed in the next chapter.

From an Arab perspective, all the governments in the region currently align with the global drive to encourage utilisation of alternative energy supplies in their energy mix. However, in numerical terms, the share of alternative energy in the Arab world energy mix is considerably less than in many parts of the world. Table 4.3 shows electricity production share from fossil fuels, nuclear and renewables in selected Arab countries. Despite relatively low numbers, the trend in most Arab countries is accelerating with major initiatives in Morocco, UAE and Jordan leading the way. The relatively low numbers can be attributed to the Arab world's massive oil and natural gas reserves providing cheap energy supplies, which create no economic incentives for the countries to diversify their energy supplies unless stricter environmental regulations are imposed.

*Table 4.3: Electricity production share from nuclear and renewables in selected Arab countries*

| Country | 2015 Share (%) | 2019 Share (%) |
|---------|----------------|----------------|
| Iraq | 3.98 | 3.03 |
| Jordan | 1.00 | 13.76 |
| Saudi Arabia | 0.04 | 0.04 |
| Syria | 2.61 | 5.17 |
| UAE | 0.25 | 3.21 |
| Egypt | 9.75 | 9.99 |
| Morocco | 15.70 | 20.66 |
| Tunisia | 3.18 | 3.70 |

Source: https://energyforhumanity.org/en/news-events/events/2020/global-energy-mix-2020-how-worried-should-we-be//
Note 1: All numbers are for 2014 and include nuclear and renewable energy

The penetration of alternative energy is gradually increasing in the Arab world, but this varies considerably between sectors (wind, solar, geothermal, biomass hydro and nuclear). At present, with the exception of the High Dam on the River Nile in Egypt as a major hydroelectric energy source, no major hydroelectric projects are planned. The Arab world's arid environment limits its potential in terms of hydro energy. In fact, water resources in the Arab world are under threat with major damming projects in Ethiopia, Turkey and Iran.

The Arab world, due to its hot and arid environment.[43,44,] possesses a huge potential to utilise its abundant solar energy, so the solar energy sector is expanding rapidly, with many projects underway. Attitudes have definitely changed and the energy mix will be altered in the next few years. The prospect of generating electricity from solar projects has gained mainstream recognition. The Ouarzazate Solar Power Station in Morocco (also called Noor Power Station) is the world's largest concentrated solar power plant,[45,46] and forms the cornerstone of rapidly expanding strategy in developing solar power in Morocco. Tunisia's renewable push unfortunately stalled, due to political instability.[47] In the Middle East several Arab countries are embracing solar power, with major projects announced or planned in Saudi Arabia, UAE, Kuwait, Jordan, and Iraq.[48] High profile initiatives include the continuous developments by Masdar in UAE and worldwide,[49] (including Al Dhafra Solar Photovoltaic (PV) plant, which is the world's largest solar power plant[50] and the NEOM project in Saudi Arabia.[51,52]

However, several grandiose mega-projects have stalled. For example, previous reports that solar power export to Europe is planned by Saudi Arabia[53] never progressed. The much-publicised plans drawn up by a consortium of European energy companies to tap into solar energy in the Sahara (the "Desertec"

---

[43] https://www.stratfor.com/analysis/bright-future-solar-power-middle-east
[44] https://www.irena.org/DocumentDownloads/factsheet/Renewable%20Energy%20in%20the%20Gulf.pdf
[45] https://www.theguardian.com/environment/2016/feb/04/morocco-to-switch-on-first-phase-of-worlds-largest-solar-plant
[46] https://en.wikipedia.org/wiki/Ouarzazate_Solar_Power_Station
[47] mees Insights, Newsletter 17/02/2021
[48] https://www.economist.com/middle-east-and-africa/2020/05/07/arab-states-are-embracing-solar-power
[49] https://masdar.ae/en/masdar-clean-energy/projects
[50] https://www.khaleejtimes.com/news/worlds-largest-solar-power-plant-set-to-operate-as-taqa-receives-uaes-financing
[51] https://www.pv-magazine.com/2020/12/02/solar-will-dominate-neom-citys-energy-mix-but-csp-may-prevail/
[52] https://www.aljazeera.net/news/politics/2020/4/26/%D9%84%D9%88%D9%85%D9%88%D9%86%D8%AF-%D9%8A%D9%88%D9%85-%D9%84%D8%A7-%D8%B4%D9%8A%D8%A1-%D9%8A%D8%B6%D9%85%D9%86-%D8%A3%D9%86-%D9%85%D8%AF%D9%8A%D9%86%D8%A9-%D8%A7%D9%84%D9%85%D8%B3%D8%AA%D9%82%D8%A8%D9%84-%D8%B3%D9%8A%D9%83%D9%88%D9%86-%D9%84%D9%87%D8%A7-%D9%85%D8%B3%D8%AA%D9%82%D8%A8%D9%84
[53] http://www.arabnews.com/news/447974

project) also hit problems.[54,55] That project aimed to produce solar-generated electricity with a vast network of power plants and transmission grids across North Africa and the Middle East. The project has stalled but remains a possibility. Although it was shelved in 2013 for technical and economic reasons[56], it was revived in 2020[57] but was then abandoned by Algeria.[58]

Although less abundant than solar, there is considerable potential for wind energy in the Arab world, both onshore and offshore. Morocco has proven offshore wind potential and has drawn plans to supply clean energy to European markets.[59] In the Middle East several projects are already planned or operational, including large operating projects in Jordan and Saudi Arabia is developing the region's largest wind farm in its north-western region.[60]

Lastly, as a result of international pressure from the USA and Israel (and nominally from the UN) nuclear energy has not been considered as a realistic energy source in many Arab countries for fear of developing military nuclear capabilities in the Middle East. However, this started to change recently with the UAE inaugurating its first nuclear power plant in 2020,[61] and Jordan initiating nuclear energy programs. Egypt and Saudi Arabia are also interested.

As a result of its growing population and aspirations for an improved quality of life, the energy demand in the Arab world, especially in the Gulf, is projected to increase[62]. This will lead to increased GHG emissions.[63] This unfortunately means that the

---

[54] Desertec Foundation, (http://www.desertec.org).

[55] http://blogs.ft.com/beyond-brics/2015/05/01/huge-solar-projects-in-north-africa-target-europe-but-will-brussels-help/

[56] http://blueandgreentomorrow.com/2013/05/16/desertec-solar-project-to-be-sidelined-says-german-broadcaster/

[57] https://ifair.eu/2020/04/03/green-hydrogen-from-the-sahara/

[58] https://northafricapost.com/43474-algeria-scraps-mega-solar-energy-desertec-project.html

[59] Mees Insights, Newsletter 10/02/2021

[60] https://www.evwind.es/2019/07/31/the-middle-easts-largest-wind-farm-will-be-constructed-in-saudi-arabia/68260

[61] https://world-nuclear-news.org/Articles/First-UAE-nuclear-reactor-reaches-full-power

[62] http://cait.wri.org/

[63] The numbers may appear decreasing in some instances if taken per capita, despite the fact that actual amounts will increase.

countries may find it difficult to reduce emissions and the limited alternative energy projects will not be enough to counter the increase.

It is clear that alternative energy sources are under-utilised in the Arab world and that there is huge potential to increase the use of those alternatives. However, a catch-22 exists here – all these alternative energy projects need capital, and since oil and natural gas revenues are the source of capital in most of the region, the countries still need oil and natural gas projects and revenue to finance these projects.

With all eyes on energy transition and significant hype generated by hydrogen as the new energy (see Section 3.6), ADNOC and Aramco are leading the blue and green hydrogen initiatives in the Arab world.[64,65] They have announced several initiatives and joint ventures which combine upstream oil and natural gas projects with solar or CCS projects. They are also exploring the potential of Ammonia as a green energy carrier. In addition, Qatar is incorporating CCS in its massive LNG expansion project, with investment exceeding US$28 billion. Qatar is currently the largest LNG producer and exporter to produce carbon neutral cargoes. It has the advantage of low-costs and thus can afford the risks of the volatility, spikes in LNG price, or the falling costs of renewables that could threaten demand.

However, while all spotlights are on the alternative energies which can facilitate the energy transition, another important factor, often under-reported, is the contribution of effective energy efficiency measures. Further implementation of energy efficiency measures could reduce energy intensity by 2-3% a year and contribute significantly to emission reduction.

With skilful management, there is the potential for the Arab world to become a leader in alternative energies. However, unlike

---

[64] https://pemedianetwork.com/hydrogen-economist/articles/strategies-trends/2021/adnoc-and-aramco-lead-gulf-nocs-blue-hydrogen-drive
[65] https://www.icis.com/explore/resources/news/2021/03/05/10614329/insight-hydrogen-may-be-big-oil-s-low-carbon-solution-in-global-energy-transition

its dominant position in oil and natural gas, it will not have the same status in this new era. Alternative energy resources are plentiful and they are available everywhere. By its nature, it is a distributed and localised market. But the Region can become a significant partner in the global alternative energy business if it decides to exploit and develop its solar and hydrogen resources, while developing the infrastructure for electricity export. Furthermore, high transport costs for hydrogen could drive its prospective users, such as steel manufacturers, to relocate around sources of supply.[66]

## 4.4   Left behind or Catching the Boat?

Following two oil price collapse episodes in six years, it is clear that the golden age of oil is drawing to a close. The facts are obvious. The world is awash with oil and natural gas and, technically, there is more than enough production potential to fulfil the world's needs. For the next few years, the average oil price will remain relatively low and will not hit the heights of 2014, rather it will be stuck in a narrow band ranging between US$50-70/bbl in real terms.[67] Incidentally, spikes of price will occur, and may hit US$100/bbl, but these instances will be short-lived.

Over the last 10 years sentiment of experts in the world has changed, transforming the conversation from peak oil supply to peak oil demand. The question being asked now is, will there be fifty years of demand?

The energy transition drive is eroding oil's share in the energy mix, especially with the push to electrify transport systems (see Chapter 5). Moreover, there is opposition in some circles, challenging the position of natural gas as a bridging fuel in the energy transition. In reality, these views are idealistic, and the role of fossil fuels will remain significant, at least till the middle

---

[66] https://pemedianetwork.com/hydrogen-economist/articles/storage-transportation/2021/hydrogen-export-costs-could-drive-users-to-supply-regions
[67] Prices nominally will reach over US$100/bbl.

of this century (See Section 3.6).

It is true that some oil and natural gas reserves will remain in the ground while the energy transition builds momentum. However, the world still needs oil and natural gas production to supply significant volumes of demand. Those countries with the most cost competitive production, are destined to benefit from their advantaged position into the mid-21$^{st}$ century and beyond.

As we saw in 2020, Saudi Arabia, in collaboration with allies within the OPEC+ framework, opened the taps, flooding the markets, aiming to control oil price and, after an initial setback, succeeded in re-establishing itself as the leader of oil markets.

However, as described earlier, the erosion of the Middle East's pivotal position in supplying world's oil needs was demonstrated when several attacks on Saudi oil installations failed to create oil price shocks and only caused a blip in the oil price. Similarly, the crisis after the American assassination of Iranian General Soleimani did not create a price shock, despite the talk of a Gulf War III, that never materialised. This is due to the fact that alternative producers are in a position to bolster production if needed.

From a strategic point of view, the Arab world lost a significant amount of its political power when the USA started shifting its interests to East Asia and the Pacific region. The dependence of the USA on Arab oil waned, in fact by 2019, it became the largest oil producer globally and a net exporter. This has been reversed following Covid-19 pandemic, but the precedent has now been established. The 2020 oil price collapse, coupled with the maturing of North American unconventional plays, resulted in lower American oil production, which is returning US dependence partially on the Middle East. This will be partly offset by the long-term structural reduction in demand, due to Covid-19, and the growing energy transition drive.

A clear picture of future oil production globally, in which some regions' outlook for future oil production is being reduced, while in the Middle East, it is being expanded. European IOCs are leading the first camp, with rapid transition away from oil and natural gas production, in favour of renewable energy as well as into utilities and energy management. The gap is being filled by the Middle Eastern countries, which see a window of opportunity for a last opportunity to expand oil and natural gas production thus filling the gap in the medium and long term. Aramco, ADNOC and Qatar Petroleum are at the forefront of companies expanding their capacity and preparing themselves for the new era.

These oil and natural gas investments in the Middle Eastern countries, most probably, will be proven to be smart. While global oil (and later natural gas) demand will shrink, as more countries embraced the energy transition, countries with cheap oil and natural gas production will be vital to supply the market needs.

For both oil and natural gas, future projects will follow emerging trends utilising innovative approaches to enhance efficiency, increasing the use of technology, improving recovery and lowering cost. There will be major drives to focus on subsea processing, digital oil field concepts, intervention maintenance, de-manning, repetition of designs, delegation to contractors and equipment suppliers, modularisation and standardisation of processes. However, these projects will be in competition with other sources to attract capital and talent. The development path will not be a simple one as it will, not only, depend on technology and economics, but also on future regulations that will play an increasingly important role in shaping the energy supply.

As I discussed earlier (see Section 4.3), the Arab states will certainly embrace renewables as a supplement besides its oil and natural gas resources. Some initiatives such as CCS may still be public relation exercises and not embraced wholeheartedly yet – but solar energy could be a game changer – capacity and generation targets are being made and the region could easily turn into a solar energy superpower. Its role will once again be a supplier of power, but it will not be a big player in solar energy manufacturing. This is a lost cause – China already dominates the market.

While we will see a transition from petrostates to electrostates and chip-states, the question that emerges is, where will Arab countries be positioned? Currently, they lack the technology to be chip-states and possess only tiny prospects of becoming electrostates. Without the mineral resources (lithium, rare earth, cobalt) and without the technological expertise or research and development (R&D) sectors, Arab countries are not ready to compete with others or develop local talent and capability in any meaningful way.

So, looking at the overall energy mix, the Arab states certainly will have a place on the boat. Alas, with the governments and regimes currently in power, they may be destined to the lower decks unless they wake up and utilise their renewable potential. They will have one leg in transition, while the other leg is rooted, fixed in oil and natural gas. It will be a delicate balancing act for the medium term, in which success is never guaranteed.

## 4.5   At What Price? Cementing of Fragile State Labelling

Having established that, despite the race to energy transition, there is still a place for sustained oil and natural gas-based economies for a while, I ask myself, if this pathway continues in the medium term, what will be the price to pay for the nations who do not rapidly embrace energy transition? The answer is obvious as we are seeing it play out in Algeria, Sudan and Lebanon. Volatile or lower oil and gas prices will play havoc in their economies, the economic deterioration will exacerbate social unrest and potentially completely destabilise some countries.

In my previous book, I researched several indicators that quantified and compared the development of civil society in the Arab world and its march towards, or retreat from, democracy with its liberal values (e.g. freedom index, democracy index, etc…).[68] Five years later, I find that almost every indicator shows no sign of improvement and, in many cases, they have deteriorated. There is however, one indicator worth examining, which is the "fragile state" index. This is because it combines inputs from many other indicators and can be used to summarise the broader situation.

We are already witnessing increased instability in the Middle East and beyond, even in countries never thought of as fragile or failed states. However, misgovernment and the Covid-19 pandemic is pushing more countries in this direction.

Just consider this extraordinary question. Is the United States of America a "failed state"? This question reverberated in American and global media in 2020 following the abysmal US government response to Covid-19 pandemic. While the question was discussed quietly by intellectuals in recent years, as the US political deadlock rendered many of its institutions ineffective and incapacitated, it is unthinkable that US mainstream media outlets would call their own country a "failed state". The term's usage in the media was once reserved exclusively to describe evil US adversaries or rogue states.

This book does not aim to discuss internal American politics. Therefore, I do not intend to engage in the mechanics of the American political debate. But it is interesting to assess the justification of the "failed state" labelling of certain Arabic states, many of which are portrayed negatively by the American media.

Technically, the term "failed state" is controversial, it is a relatively recent expression and can carry significant geopolitical consequences. There is no consensus on a definition and the term, introduced by scholars in mid-1990s. It is used by many researchers and journalists describing the effectiveness of the government to determine if a state is failed or not. However,

---

[68] Basel Nashat Asmar, Fossil Fuels in the Arab World: Seasons Reversed – Oil and Politics Interplay in the Arab World, 2050 Consulting, London, UK, 2017

despite the term's relatively recent introduction, the concept of "state failure" as an actual phenomenon, has been debated extensively for a long time and depending on your interpretation, has been part of the political reality for decades, if not centuries. Definitions of the "failed state" term vary, but, are always described in contrast to the successful functional states that are judged to be the norm. The Oxford dictionary defines the term simply *"a state whose political or economic system has become so weak that the government is no longer in control"*.

The Fund for Peace[69] introduced a quantitative tool to measure the categorisation of failed states. It lists detailed criteria of what constitutes a failed state. These include loss of control of its territory; the erosion of legitimate authority to make collective decisions; the state's inability to provide public services and its inability to interact with other states as a full member of the international community.

Since 2005 the Fund for Peace produced the Failed States Index (FSI)[70]. The index was later renamed Fragile States Index (FSI) in 2014. This was attributed to scholarly criticism of the term "failed", where some claim it is the incorrect terminology. Because the term is in the past tense, they claimed it suggests and implies that failure is everlasting and there is no way back.

Since its inception, FSI was published jointly by the Fund for Peace with the Foreign Policy Magazine until 2018 and then with the New Humanitarian[71]. The Index scores each country, based on twelve social, economic, political and military indicators. It spans an overall score between 0 and 120. It then categorises states into several categories, which evolved over time and are currently ranging from "very high alert", in the bottom category, to "very high sustainable", in the top category.

---

[69] The Fund for Peace is a non-governmental research and educational institution, which researches ways to prevent violent conflict and promote sustainable security. It was founded in 1957, and is based in Washington DC. (www.fundforpeace.org).
[70] https://fragilestatesindex.org
[71] https://www.thenewhumanitarian.org/

In my view, the renaming is pitiful, and the exercise is a purely cosmetic change. Using "fragile", instead of "failed", follows a growing trend of unnecessary sensitivity spreading in the scholarly community. All indices or classifications, including the ones discussed in this book, are increasingly adopting a "politically correct" attitude, where they tend to avoid naming things with their correct names, but rather mask the true terminology to make it appear less offensive.

With the term failed or fragile state clarified, let's examine the status of the Arab world, in terms of the FSI, focusing on the last five years. Figure 4.7 below shows the change in scores for the Arab countries between 2015 and 2019, as well as their ranking and classification in 2019. Note that the ranking scores mean that, the lowest number indicates a very high fragile state.

At the dawn of the new millennium, the Arab world governance was entirely authoritarian, where regimes ruled with an iron-fist, but maintained relatively functional services and effective control. The exception was Somalia which was a disintegrating failed state with no functional government. Then the Anglo-American invasion of Iraq in 2003 turned Iraq into a failed state. Prior to the invasion, the Iraqi government was authoritarian, but it enjoyed effective control of its territory and, despite harsh sanctions, provided adequate health and education services to its citizens. As a result of the war and the mistakes made governing the country by the coalition, the country's governance structure and institutions collapsed, plunging the country into chaos.

The aftermath of the invasion and its consequences triggered economic and political instability, creating a domino effect of instability across the region. This included the uprisings known as the Arab Spring and then the Arab Winter. As a result, the number of fragile states in the Arab world increased initially to five, reaching a peak of eight in 2015.

In 2019, out of 31 countries falling in the "alert" categories, thus labelled as "fragile", seven are members of the Arab League.

These countries are Yemen, Somalia, Syria, Sudan, Iraq, Libya and Mauritania. Since 2010, another Arab country, namely Egypt, the largest member of the Arab League had this status for five years and continues to teeter on the brink. Should Egypt fail completely, it could have catastrophic results at home and calamitous regional consequences. Djibouti is also on the threshold. In 2019, Yemen acquired the dubious honour of top of the table, replacing South Sudan. Lebanon is also spiralling rapidly into this status.

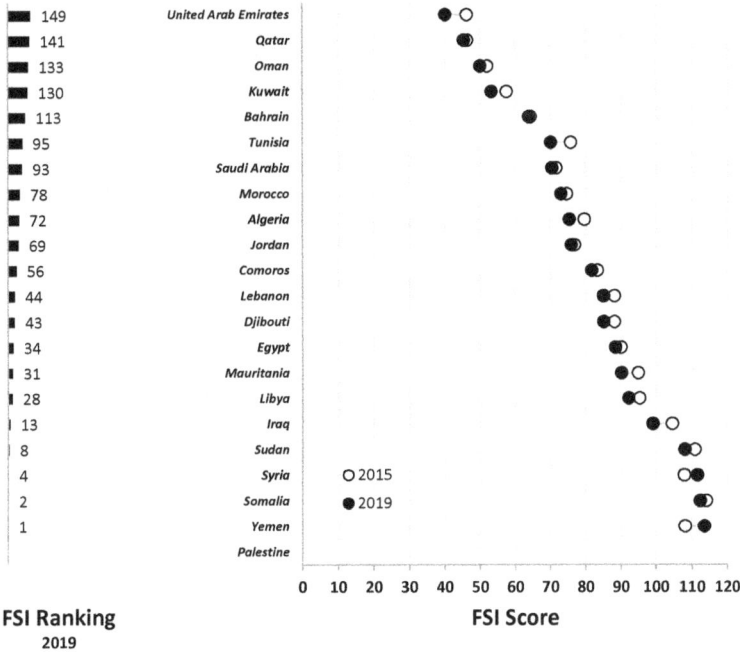

| FSI Ranking 2019 | | FSI Score |
|---|---|---|
| 149 | United Arab Emirates | |
| 141 | Qatar | |
| 133 | Oman | |
| 130 | Kuwait | |
| 113 | Bahrain | |
| 95 | Tunisia | |
| 93 | Saudi Arabia | |
| 78 | Morocco | |
| 72 | Algeria | |
| 69 | Jordan | |
| 56 | Comoros | |
| 44 | Lebanon | |
| 43 | Djibouti | |
| 34 | Egypt | |
| 31 | Mauritania | |
| 28 | Libya | |
| 13 | Iraq | |
| 8 | Sudan | |
| 4 | Syria | ○ 2015 |
| 2 | Somalia | ● 2019 |
| 1 | Yemen | |
| | Palestine | |

*Figure 4.7: Comparison of Fragile States Index (FSI) in the Arab countries between 2015 and 2019*

Source: The Fund for Peace
Note 1: In 2015, scores are interpreted as <30 Sustainable, 30 - 69.9 Stable, 70 – 89.9 Warning, 90 – 120 Alert. In 2019, scores are interpreted as
Note 2: Palestine has no independent score published. Its data are incorporated into Israel's score.

The information from Figure 4.7 reveals that the number of failed states has rapidly increased in the region. It shows a surprising variety of scores, where the Gulf countries appear to be beacons of stability, contrasting with what is happening elsewhere in the region. The scores of both Syria and Libya unsurprisingly increased sharply, pushing them deep into failed state territory. No doubt, Yemen's future score will soon place it on the top-ten list, following the brutal civil war there. Furthermore, we can see that the trend is continuing, with most Arab countries in the region moving to the right of the graph. The results demonstrate the complete failure of intra-Arab policy and the numerous blunders committed by the Arab countries dealing with each other. They also indicate how Western foreign policy is floundering in the Arab world.

However, eventual stability, especially with diminishing oil and natural gas revenues, may be under threat even in the Gulf. With tightening budgets and spending cuts, the public will become more outraged about their countries' rife corruption, which could easily lead to instability. In addition, imposing taxes, especially in GCC (Saudi tripling its VAT, Oman introducing new income tax) may cause social unrest. All this may lead to fragility!

Figure 4.8 shows the average Arab score, compared to selected countries and regions, here the Arab score reported is a weighted average, with calculations based on population. The international standing of the Arab world, although numerically improved slightly, has deteriorated, relative to the group of countries and regions benchmarked, except against the United States, Brazil and Turkey, which are controlled by populist presidents.[72]

The results reflect the effects of corruption and repression in the region. Achieving a failed state score is the culmination of many factors including corruption, separatism, authoritarianism etc., as they are the building blocks contributing to that collapse.

Alarmingly, the factors causing increasing numbers of failed states in the Arab world, appear to be spreading. It is also spreading affecting neighbouring countries, which are either already failed states (e.g. South Sudan,[73] Central African

---

[72] Refers to Trump's administration in the US.
[73] South Sudan gained independence from Sudan, which was already a failed state. Thus, it started

Republic, DR Congo) or on the brink of failing (e.g. Mali). The data indicates that the phenomenon is not limited to the Middle East is creeping into other regions, e.g. Ukraine, Georgia and, to some extent, Greece. The instability appears to have reached even to Israel, which after four elections in two years continues to suffer from political instability.

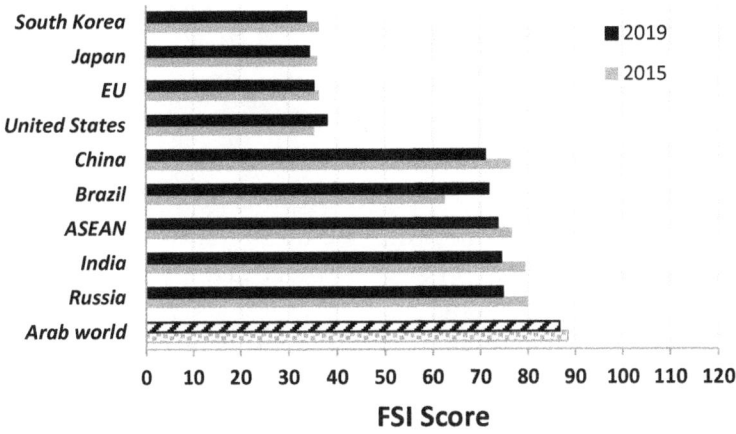

*(a) Relative to selected countries and regions*

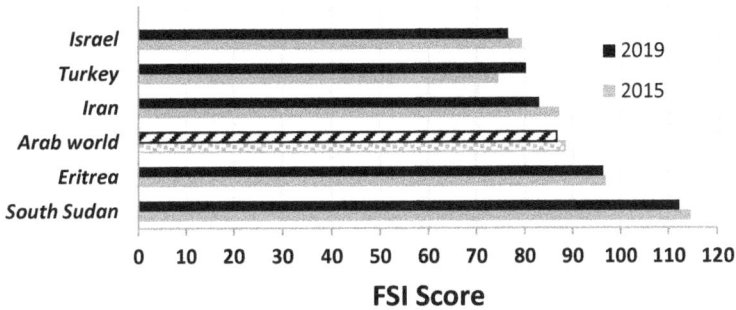

*(b) Relative to neighbouring countries*

*Figure 4.8: Comparison of Fragile States Index (FSI) between the Arab countries and selected countries in 2019*

Source: The Fund for Peace
Note 1: Israel number reported incorporates the West Bank of the Palestinian Authority.

its own existence as a failed state.

Between 2015 and 2019, as most civil unrest subsided, there was a tangible, albeit slight, improvement in the Arab world stability conditions. However, further unrest started in 2019 in several countries and, despite restrictions imposed during the Covid-19 pandemic, tensions continue to simmer and could escalate with the triggers described.

Is it correct to use the FSI as a measuring index? Some scholars argue that this index is flawed or biased[74] somehow, as it conceals other significant negative aspects, i.e. corruption and lack of democracy. In effect it measures a state's strength, stagnation and stability. It rewards wealthy nations, even if their scores in other indicators are low. Others argue that the true measure of a nation is not the number and magnitude of challenges it faces, but how it rises to meet them, but in most of the countries topping the ranking (i.e. failed states) the truth is they are unable to meet any of the challenges whatsoever!

Recently, there have been suggestions that, in this globalised world, not all states will, or should, survive in their current form. The population of many failed states might benefit more from living indefinitely in a "non-civil" society, than in a dysfunctional state artificially sustained by international efforts.[75] Some observers suggest that the governance of the failed states should transfer to international trusteeship.[76] While this initially sounds questionable, recent events proved that it can be possible. The population of Somalia survived for over twenty years without a functioning "state".

Closer to home, the political impasse in Lebanon did not stop the country functioning until 2019 – but it reached its limits and exploded since. While Israel continues to function despite impossible political impasse.

Certain countries continue to demonstrate that politicians can be

---

[74] Edward Said wrote about the way the West describes and portrays the East – it might be relevant to clearly acknowledge the inherent Eurocentric bias of the FSI.
[75] Failed States, or the State as Failure?, Rosa Ehrenreich Brooks 72 (4), 1159 - 1196, 2005
[76] https://www.icrc.org/eng/resources/documents/misc/57jq6u.htm

surplus to requirements and such countries can function with civil servants running the show. Belgium functioned with a caretaker administration, without an elected government, for almost 600 days in 2010 and 2011. It has fared relatively better than its peers combating the Covid-19 pandemic, despite being governed by caretaker government. This proved that Belgium, which has long been written off as a dysfunctional state, with its civil servant caretakers was able to devise a pandemic response that has been remarkably functional.[77] Furthermore, Taiwan, unrecognised by many and deprived of WHO membership, showed one of the most effective responses in combating Covid-19.

Others show that demagogic politicians do more harm – consider the ignominious world leaders Boris Johnson (and Brexit), Bolsonaro, Trump, Modi and Sisi – whose governments are in denial and who routinely manipulate facts and political information.[78,79,80] Examine their responses to the Covid-19 pandemic and the death toll levied on their populations. It is interesting to observe that, in response to Covid-19, authoritarian regimes were highly effective, i.e. China and Vietnam – but Korea disapproves. While I am not in favour of tyrannical dictatorships or authoritarian regimes, repressed citizens obeyed their government restrictions and their societies bore much less impact from the pandemic than many First world democracies.

While I have laid bare the unpalatable facts, inviting the reader to judge for themselves, what the future may hold for the Arab world, my analysis shows that unless there is miraculous change, it is on a path to 'less significance' rather than irrelevance. Without miraculous change, that will have to be good enough.

---

[77] https://www.aljazeera.com/indepth/opinion/failed-state-managed-coronavirus-outbreak-200413152555554.html

[78] https://midan.aljazeera.net/reality/politics/2020/5/1/%D9%83%D9%88%D8%B1%D9%88%D9%86%D8%A7-%D9%88%D9%86%D8%B8%D8%A7%D9%85-%D8%A7%D9%84%D8%B3%D9%8A%D8%B3%D9%8A

[79] The Economist 04/11/2020, p 5

[80] https://www.theguardian.com/technology/2020/apr/02/twitter-accounts-deleted-linked-saudi-arabia-serbia-egypt-governments

## Chapter 5
# *STALLED IN THE*
# *TRANSPORTATION REVOLUTION*

The Covid-19 outbreak, like nothing else, is a defining moment demonstrating the world's dependence on oil markets for transport. It has also clearly shown, how the survival of the oil market relies on the freedom of the population to travel and driving their vehicles, without restrictions. The lockdowns imposed by most countries on their populations, effectively banned people from driving their cars, sent shockwaves throughout the oil markets, as it effectively destroyed demand. The implications of that demand destruction have already been discussed in the previous chapters.

With this dependent relationship so starkly evident, one would expect that petrostates, whose economies depend on expanding oil demand, would pay more attention to personal mobility, driving trends and the behaviours that will affect the future direction of the transport industry. However, scanning the various groups involved in the fast-transforming sector, it is evident that the opposite is happening. Unbelievably, there is minimal investment from oil producing countries in the main groups of the transportation sector and even less effort in establishing or promoting local developers. Interestingly, all the major groups are in oil consuming nations, which aspire to cut their oil dependency and reliance.

It is important to set the context for this chapter, which will touch only on certain aspects of transportation. At the moment, media outlets are increasingly using the term "mobility" to refer to transportation. It is essential to understand the meanings of the two terms.

In simple terms, the word **mobility** is defined as the ability to move easily or be moved freely and easily. The movement can be from one place, social class or job to another. It has different meanings, depending on the context in which it is used and therefore it is interpreted differently in each sector. This can be clearly seen in Wikipedia's disambiguation page for mobility, where over a dozen of associations with the word are mentioned.[1]

On the other hand, **transport**, or **transportation**, is the movement of humans, animals and goods from one location to another. Modes of transport include air, land (rail and road), water, cable, pipeline and space. Overall, transport systems encompass infrastructure, vehicles, services and operations.

Looking at the terms more broadly, mobility is like an umbrella term that includes transportation, as it encompasses technologies and services that enable people and goods to move around more freely.

In this book, I am not going to use mobility, as synonymous with transportation, since mobility in fact means "access to transportation". For the rest of this chapter, I will focus on selected aspects of transportation that have significant impacts on fossil fuel demand. Without mobility, transportation is meaningless. Improving people's mobility should be the goal of any transportation project. Three ingredients are essential for any transportation project to succeed: affordability, safety and time (journey duration).

Due to rapid technological advances in the transportation sector, it seems that the two transportation and software sectors are becoming increasingly amalgamated, with software rapidly replacing many transport functions that previously involved manual or mechanical efforts. The sectors overlap even more when it comes to services within the transportation, with many transport service companies now regarded as IT companies providing IT platforms.

---

[1] https://en.wikipedia.org/wiki/Mobility

It is outside the scope of this chapter to address the media and political agenda around climate change or the 2° scenario, the two issues are relevant to the subject of transportation (see Section 3.6 for a brief discussion). Note that, although recent international government policies aim to reduce fossil fuels emissions, there is a significant difference between sectors, as oil is dominant in some sectors, while natural gas is dominant in others. Oil is a transportation-dominant fuel, whereas natural gas is used for power generation and other uses in factories and so on.[2] The latter fuel's effect on transportation will increase with the electrification of the sector.

Here I am looking at two aspects which will be significant: mobility and telecommunications, with their implications on technology, work from home, etc. By examining selected topics in the transportation sector, we can see how these developments are changing the way we live. The changes will have massive impact on fossil fuel consumption and the Arab world. I also try to identify any Arab world influence and track any contribution that might exist.

## 5.1    Transportation Energy Consumption

According to the EIA's classification, energy is viewed as either primary, which is an energy form found in nature that has not been subjected to any human engineered conversion process, or secondary, which is a carrier of energy, produced by conversion from a primary energy source. Primary energy sources include fossil fuels (petroleum, natural gas, and coal), nuclear energy, and renewable sources of energy. Electricity is a secondary energy source that is produced (generated) from primary energy sources.

Besides industrial, commercial and residential sectors, the transportation sector is the fourth sector of so-called "end-use" energy sectors. It consumes primary energy directly, as well as

---

[2] https://www.spglobal.com/en/research-insights/articles/insight-conversation-saad-al-kaabi-minister-of-state-for-energy-affairs-qatar

electricity produced by the electric power sector.

In the last ten years, energy consumption of transportation sector increased in physical terms, but its share of overall consumption remained flat. Figures 5.1 and 5.2 illustrate this.

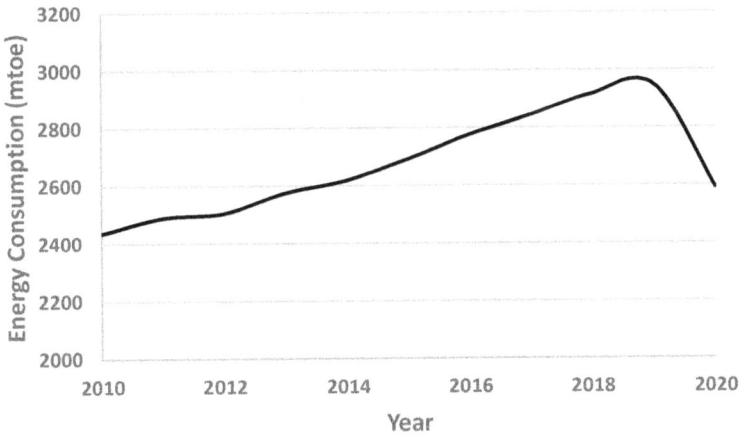

*Figure 5.1: Transport energy consumption*

Source: IEA

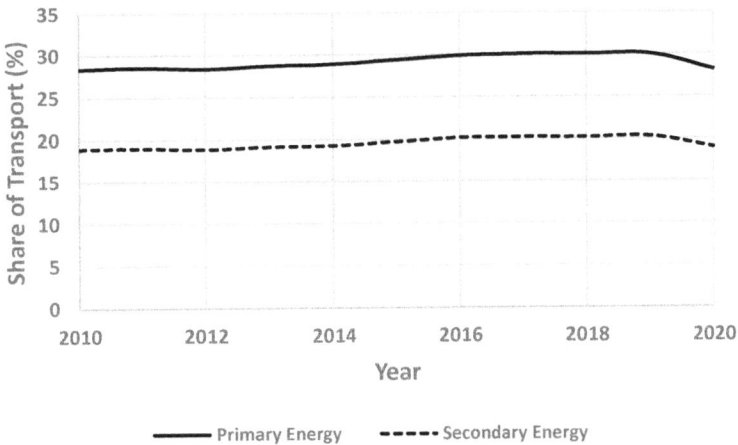

*Figure 5.2: Transport share of primary and secondary energy consumption*

Source: IEA

As I have discussed in the previous chapters, the mix of energy sources is changing, with energy transition reducing the share (albeit not the absolute quantity) of fossil fuels. However, this transition is not equal in all end-use sectors, where different trajectories are being followed. See Section 3.6 for more details on energy transition.

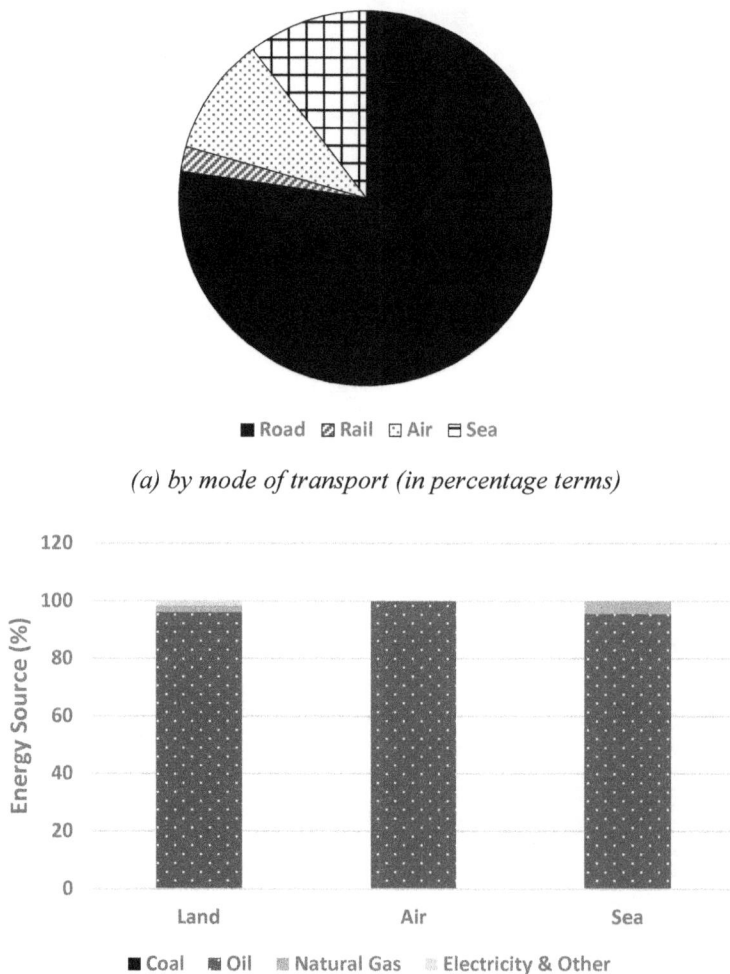

■ Road  ☑ Rail  ⊡ Air  ⊟ Sea

*(a) by mode of transport (in percentage terms)*

■ Coal  ▦ Oil  ▦ Natural Gas  ▥ Electricity & Other

*(b) by energy source*

*Figure 5.3: Transportation sector energy consumption by transport mode and energy source*

Source: IEA

Figure 5.3 below shows the consumption of transportation sector by both energy source and transport mode. It is clear from the graph that land transport (road and rail) dominates the sector consumption and oil is the dominant energy source used.

From the graph, we can see that rail transport constitutes less than 3% of land transport energy consumption, despite it being more efficient way of transport.[3] The majority of consumption is guzzled by land transport. The facts above illustrate the co-dependency between road transport and oil. It also signifies that to reduce oil consumption, a transformative structural change is needed within the road transport sector. At the moment, we are in the midst of this transformation and I will be discussing this in the next few sections.

In the last few years (pre-Covid-19 pandemic), the growth in air travel climbed explosively, with the introduction of many new budget airlines, offering cheap travel. Air travel overtook water transport as the second ranked energy consumer in the transport sector. However, in energy terms, air travel is not efficient compared to water transport, which is considered to be the most efficient way transporting freight globally.[4] Nevertheless, water transport mode is totally dominated by freight and passengers' proportion in it is miniscule.

Despite all the inefficiencies, air transport for goods, which is more expensive, with volume limitations, has one distinct advantage, which is journey time. Also, it has a much larger range. Sea or ocean-going transport requires costal ports and accessible navigable routes to go inland but we can fly almost anywhere on the face of the planet. In the Covid-19 outbreak, this proved invaluable as medical supplies had to be air transported.

---

[3] https://ibir.deutschebahn.com/ib2016/en/group-management-report/group-performance-environmental-dimension/progress-in-climate-protection/energy-efficiency-increased/
[4] https://ibir.deutschebahn.com/ib2016/en/group-management-report/group-performance-environmental-dimension/progress-in-climate-protection/energy-efficiency-increased

## 5.2    Road Transport

In the last five years, land transport has made its way up the list to the forefront of the public agenda, with important discussions involving public health (pollution), the environment (climate change), infrastructure provision and cost, as well as convenience. Balancing the needs and costs of public transport versus those for private transport usage is a major debating point in political circles, triggering passionate arguments with many diverse, points of view.

In the minds of the public, especially in the US, road transport mainly equates to cars. In several countries, other types of micro-mobility vehicles (e.g. bikes, e-scooters) compete with cars and, where the public are encouraged to walk or cycle, cycling friendly policies are gaining momentum.

The increasing awareness of the effects of climate change has been shaping popular dialogue over the last five years. The public are increasingly in favour of limiting emissions generated by the transport system, especially diesel and gasoline vehicles, endorsing a shift into a greener system dominated by electric cars. This is discussed in detail later in Section 5.2.1.

Pre-Covid-19, most governments were promoting public transport, as mass transit, whether trains or buses, is more efficient than cars. They argued that these modes are capable moving large numbers of people efficiently, i.e. with less fuel consumption, resulting in lower impacts on the environment, especially when compared to car usage, where each car often carries a single occupant.

Reflecting on the car ownership model of the 20th century, it is clear that the old model is flawed. Cars are the least used expensive items that people buy. They sit idle for approximately 95% of the time. Apart from the initial outlay required to purchase their car, the rapidly depreciating capital cost, owners then incur numerous additional operating costs including fuel, parking, annual taxes, insurance, MOTs, maintenance etc. Consider, how would you feel if, every time you paid for your groceries, you only used 5% of them? Or having purchased the most up-to-date expensive smartphone, you only charged it for 5% of its capacity? Would you reconsider the wisdom of such purchases or find a more efficient alternative?

All these habitual purchasing and behavioural trends were challenged during the initial months of the Covid-19 pandemic, then again when lockdown and its severe restrictions were being eased. These peculiar situations exposed a major flaw in the promotion of public transport. Crowding passengers onto train carriages and bus decks, like sardines in tins, is not viable when public health must be maintained.

Public health guidelines recommended during the pandemic, encouraged people to work from home where possible, with those who were required to continue travelling to their workplace, to socially distance while commuting. Social distancing measures, even when so many workers were not commuting, reduces the capacity of public transport by over 80%, rendering it unfeasible economically. In these circumstances, many European transport ministers advised the public to reduce exposure to the virus by avoiding public transport and using private cars instead, contradicting the message they preached for years to the contrary.

This promotion of private car usage is an extraordinary blip, an interruption in an otherwise continuing shift in attitudes towards car usage. In many countries, attitudes are changing, and perceptions of car ownership are being relegated from desirable to unnecessary, maybe even unethical. There are several factors that have led to this discouragement of car ownership.

Firstly, individuals are making less car journeys. They are increasingly using delivery services to shop for food, restaurant meals, household goods, luxury purchases and other amenities. Amazon's share price increased by 67% in 2020, valuing it at over US\$1.5 trillion[5] and cementing the position of its owner Jeff Bezos as the richest man in the world. His wealth is estimated to be over US\$190 billion by the end of 2020 and putting him on a trajectory to be the world's first trillionaire.[6]

Secondly, the increase in ride-hailing and ridesharing services has made owning a car an unnecessary luxury. These relatively cheap and convenient services now threaten the automotive construction industry, as the services are much more economical than car ownership, for most urban dwellers.

The above services are now provided by a different type of company that define a new sector, often referred to as MaaS - mobility-as-a-service. This is discussed in detail in Section 5.2.2.

The lockdown work and travel restrictions during the Covid-19 pandemic also proved that many jobs can be performed from home, in many cases with increased productivity, thus demonstrating that commuting to work can be reduced.[7] The implications of this will be enormous, reducing considerably the number of trips taken by individuals, thus affecting both land transport, and international business travel using air transport. While these changes to the work landscape were theoretically possible pre-Covid-19, the lockdown was a real-time "proof of concept" now called by many "the mobility revolution". We can be certain that this paradigm shift will have major impacts on automotive and oil demand. One interesting argument defines this phenomenon as the "three zeros", i.e. that the future of the automotive industry will be zero emissions, zero accidents and zero ownership.[8]

---

[5] https://www.google.com/search?q=amazon+share+price&oq=amazon
[6] https://www.businessinsider.com/jeff-bezos-on-track-to-become-trillionaire-by-2026-2020-5?r=US&IR=T
[7] Other benefits of this paradigm change for the workers include saving time spent commuting, cost of commuting and additional living expenses related to work, e.g. snacks, meals and drinks. Other un-costed benefits are also improved quality of life, improved work/life balance and better health and familial relationships.
[8] The Mobility Revolution: Zero Emissions, Zero Accidents, Zero Ownership Kindle Edition by Lukas Neckermann

The shrinking share of car transport is allowing other modes of road transport to prosper. Those modes are expanding, albeit at different growth rate and, overall, their share of the road transport market remains small. These micro-mobility modes include motorbikes, mopeds, electric motorcycles, scooters, bicycles, motorised bicycles, e-scooters, kick scooters, personal transporters and, last but not least, walking, etc. The change in attitude will result in reshaping roads, with more closures, more cycling lanes and more pedestrianised streets.

It is evident that the traditional automotive industry is in flux. We are in the midst of the emergence of a new mobility ecosystem, which is being shaped by immense innovation.[9] New entrants with disruptive business models are emerging, and will certainly transform the land transport sector beyond recognition.

As the "mobility revolution" continues, we will see the viability of many rail services and stations will be challenged. Moreover, with the exception of several busy routes, public rail transport is often subsidised,[10]. Will governments be prepared to continue subsidising rail services that are unprofitable? Rail electrification is expanding (including battery-electric locomotive) but it is not discussed here.

In the next two sections I will discuss road transport in more detail, focusing on the transformation of vehicle manufacturing and its fusing into becoming a service sector. As glaringly demonstrated, by the effects of reduced driving on oil demand, and thus oil markets, oil price and oil future. It is essential to understand the state of these industries briefly, as it has significant implications on the future of the oil industry and, by association, Arab petrostates.

### 5.2.1 Vehicles Manufacturing Evolution

The history of the automobile manufacturing is a fascinating tale.

---

[9] https://www.ey.com/en_gl/automotive-transportation-future-mobility
[10] https://en.wikipedia.org/wiki/Rail_transport

It demonstrates humans endeavour to invent and excel, to tame nature's forces and utilise them to enhance their standards of living and extend their reach.[11] For millennia humanity depended on domesticated animals or human power for land transport. The concept of motorised transport made an appearance in the eighteenth century and its development changed the way we live. That history is outside the scope of this book, but it is worth considering that it is estimated to be over 100,000 patents created for the modern automobiles and motorcycles.[12]

The latter part of the nineteenth century saw several attempts to commercialise motorised vehicles in Europe and the US. Initially, for decades, steam, electric and petrol/gasoline-powered vehicles competed with each other until gasoline internal combustion engines achieved dominance in the 1910s. This dominance (and the start of the oil century) coincided with the rise of liquids as fuels in the combustion engines, which were initially hampered by lack of suitable fuels and often a mixture of gases was used instead.

The ubiquitous petrol combustion engine prevailed in the twentieth century. Its advancement formed the core of an expanding automotive industry that prospered after World War II, evolving into a global industry with increasing standardisation, platform sharing, computer-aided design and increased use of electronics for both engine management and entertainment systems. Several companies now dominate the industry, establishing factories in multiple locations, across borders, to reduce costs and development time.

Concerns about the environmental impacts of the combustion engine and rising fuel costs, started to appear late in the twentieth century, leading to the emergence of electric cars as an alternative. Now, in the early twenty first century, it seems that the electrical engine will make the combustion engine obsolete.

---

[11] See https://en.wikipedia.org/wiki/History_of_the_automobile
[12] https://en.wikipedia.org/wiki/Car

While researching this subject, I was overwhelmed by the number of acronyms used and the alternatives to combustion engine solutions available on the market. I have included a few paragraphs below as a summarised crash course for the similarly unfamiliar reader.

Firstly, to anyone not aware of the fundamental difference between the two types of engines, here it is. Combustion, also known as burning, is the basic chemical process of releasing energy from a fuel and air mixture. It is an exothermic chemical reaction that involves fuel and oxidants. There are two types of combustion engines: internal and external. In an internal combustion engine, the ignition and combustion of the fuel occurs within the engine itself. The engine then partially converts the energy from the combustion to usable work. An external combustion engine (e.g., steam engine), on the other hand, is a heat engine where a working fluid, contained internally, is heated by combustion in an external source, through the engine wall or a heat exchanger. The fluid then, by expanding and acting on the mechanism of the engine, produces motion and usable work.[13] In both types of combustion engines, heat produced as a result of combustion and a few dozen parts working together, is used to propel a car from its stationary position. The concept of the electric engine revolves around magnetism. The engine draws power from the battery, creating a magnetic force that propels the car forward. The three primary components of an electric car are the electric motor, controller and battery.[14]

Nowadays, the scene is complex, with a mix of engine types and fuel types on the road. With the growth of electric vehicles, numerous types of hybrid vehicles and new combustion engine vehicles, using alternative fuels, have numerous acronyms that are used to describe them. The following acronyms are essential to understand the current state of the industry and be able to follow its developments:

---

[13] The steam engine falls under these criteria.
[14] http://www.suntecautoglass.com/blog/2018/07/03/electric-vs-combustion-engine-what-are-the-differences.html

- ICE vehicles – Internal combustion engine vehicles, which are the traditional combustion engine cars, powered by gasoline, diesel, biofuels, LPG, natural gas (compressed natural gas (CNG) or liquefied natural gas (LNG)).

- EV – Electric Vehicles, which are subdivided into several categories:

  o HEV – Hybrid electric vehicles, which are powered by a combination of an ICE and an electric motor (hybrid vehicle). Either or both the ICE and electric motor are in operation. The batteries can be charged in different ways, by spinning an electric generator when the ICE is operating, or by converting the vehicle's kinetic energy into electric energy through systems like turbochargers and regenerative brakes.[15]

  o MHEV – Mild hybrid electric vehicles, which are similar to a HEVs, but their electric motors are not powerful enough to power the car on their own and can only assist the ICE. They do not have an electric-only mode of propulsion.

  o PHEV – Plug-in hybrid electric vehicles, which are similar to HEVs, but they differ by having their rechargeable battery that can be charged by plugging into a power source, in addition to regular hybrid charging alternatives.

  o BEV – Battery electric vehicles, which have no internal combustion engine or fuel tank at all and run on a fully electric drivetrain. They are powered by rechargeable batteries that need to be plugged in to a power source to charge and, depending on the vehicle and battery capacity, they have varying charging times and driving ranges.

---

[15] https://www.canadianfuels.ca/Blog/2016/September-2016/ICE-HEV-PHEV-and-BEV-%E2%80%93-What-they-mean-and-what-s-under-the-hood/

- o NEV[16] – Neighbourhood electric vehicle, which is a US designation for battery electric vehicles that are usually built to have a top speed of 25 miles per hour (40 km/h) and have a maximum loaded weight of 3,000 lb (1,400 kg). The non-electric version of the NEV is the motorised quadricycle.

- FCV – Fuel cell vehicle, or fuel cell electric vehicle (FCEV), is a type of electric vehicle which, instead of a battery, uses a fuel cell, or works in combination with a battery or supercapacitor, to power its on-board electric motor. Fuel cells in vehicles generate electricity to power the motor, generally using oxygen from the air and compressed hydrogen.

One main difference between the ICEs and electric motors is that the latter are inherently flexible. EVs can have multiple motors that can work in tandem or power up each of the four wheels independently. This gives the ability to offer a variety of power capacities (sizes). A traditional ICE manufacturer would be required to incur the cost of engineering and building many more engine sizes.[17]

There are a variety of batteries used in EVs. These include lithium ion, molten salt, zinc-air, and various nickel-based designs.

EVs are being promoted as the automotive solution to reduce pollution. They are advertised as environmentally friendly, resulting in low or no carbon emission, depending on the type of EV. It is also claimed that they provide higher fuel economy, are smoother to drive, being quieter, with reduced engine sound, requiring lower maintenance and offering the convenience of charging them at home.

---

[16] Note that in China NEVs is the abbreviation used for new energy vehicles, which refers to vehicles that are partially or fully powered by electricity such as PHEVs and BEVs.
[17] https://insideevs.com/features/342865/four-advantages-of-manufacturing-evs-over-ice-vehicles/

While some of these characteristics are true advantages, others are questionable. For example, while EVs certainly cut engine emissions, emissions are effectively shifted to electric plants and pollution resulting from the mining required to extract essential materials for their fabrication. Batteries especially can be very damaging and without recycling, electric vehicle batteries could lead to vast amounts of highly toxic waste.[18] Also, despite the fundamental technology – burning fuel to create power – remaining the same for decades, the vast improvements in the ICE efficiency have reduced fuel usage enormously and thus have lowered their emissions,[19] In the last few years the impact of this increased engine efficiency was responsible for more oil demand reduction. EVs are increasingly displacing petroleum consumption and were responsible for 3% reduction in 2019. According to Bloomberg NEF long-term EV outlook, which forecast that by 2040, EVs could displace as much as 6.4 million barrels a day of oil demand, while fuel efficiency improvements in ICEs will erase another 7.5 million barrels a day.[20] A significant portion of the reduction will be from buses and medium and heavy vehicles rather than passenger cars. Just to put it in context, for every 1000 electric buses on the road, 500 barrels of diesel are displaced each day, while 1000 BEVs remove just 15 barrels of oil demand.[21] Thus at current conditions, more oil demand is saved by greater efficiency in combustion engines, than in transitioning to electric vehicles. However, the aftermath of Covid-19 outbreak, changing attitudes may tip the balance in EVs direction.

In order to promote EVs, positive claims are always made in literature and media, stating that "An electric vehicle converts over 50% of the electrical energy from the grid to power at the wheels, whereas the gas-powered vehicle only manages to convert about 17%–21% of the energy stored in gasoline."[22] Or

---

[18] https://www.theverge.com/2019/11/6/20951807/electric-vehicles-battery-recycling

[19] See for example https://www.autonews.com/article/20181112/OEM06/181119968/amid-ev-hype-the-internal-combustion-engine-keeps-improving

[20] https://www.bloomberg.com/news/articles/2019-03-19/how-much-oil-is-displaced-by-electric-vehicles-not-much-so-far

[21] https://about.newenergyfinance.com/blog/forget-tesla-chinas-e-buses-denting-oil-demand/

[22] https://www.alliedmarketresearch.com/electric-vehicle-market

"EVs convert over 59-62% of grid energy to the wheels. Conventional gasoline vehicles convert only some 17%–21%."[23] While the numbers may appear true, once losses from generating electricity are taken into account, (often from fossil fuels), the advantages of EVs diminishes considerably. This does not mean that EVs do not make a difference, but the magnitude is not as advantageous as we are being led to believe.

In fact, EVs have several drawbacks compared to ICE vehicles. The main two are the shorter driving range - although ranges are improving, and the recharge time, which is often takes hours compared to the few minutes required to fill an ICE vehicle fuel's tank. Furthermore, the infrastructure for charging EVs is still lacking compared to that available for ICE vehicles.

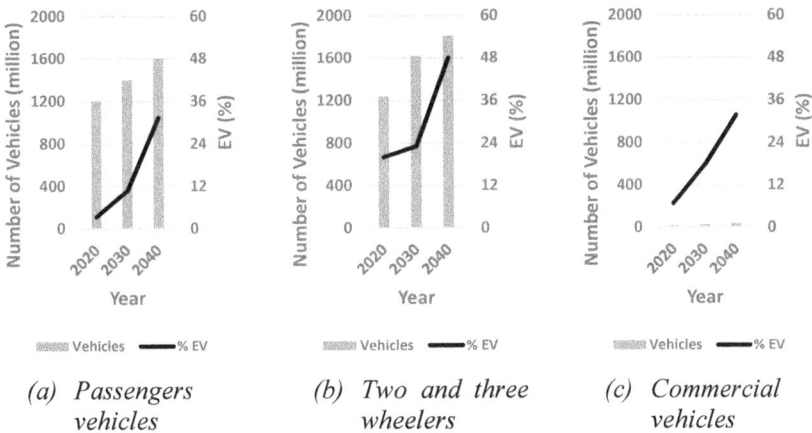

| (a) Passengers vehicles | (b) Two and three wheelers | (c) Commercial vehicles |

Figure 5.4: World vehicle population
Source: https://bnef.turtl.co/story/evo-2020
Note 1: ICE numbers include diesel, gasoline, and natural gas engine vehicles.
Note 2: EV numbers include FCV.
Note 3: Commercial vehicles include light, medium and heavy-duty vehicles plus buses.

At the beginning of 2021, the car market was still dominated by ICE vehicles, but the EV market have been growing massively. Figure 5.4 shows the world's vehicle population and the proportion of EVs. As can be seen in the graph, the share of EVs

---

[23] https://www.fueleconomy.gov/feg/evtech.shtml

is increasing considerably and many manufacturers are announcing plans to eliminate ICEs in passenger cars. GM has announced its intention to eliminate combustion engines by 2035.[24] Despite this, ICEs are still expected to dominate the fleet numbers (including cars and commercial vehicles) in 2040.

According to Allied Market Research, the global electric vehicle market value was valued at US$162.34 billion in 2019 and is projected to reach US$802.81 billion by 2027.[25] Others put the value much higher, exceeding US$3.1 trillion in 2027.[26] As a reference, the global commercial vehicles market size was valued at US$1.51 Trillion in 2019 and is expected to reach US$2.55 trillion by 2027.[27]

The environmental credentials that favour EVs are becoming increasingly important as a driving factor growing their market share. Governments continuous tightening of emission regulations and the introduction of increasingly stringent targets, such as aiming for restricting vehicles sales to "zero emission models only" by set targets in order to conform with climate change aspirations and reduce the usage of ICE vehicles, the public are going to be pushed hard to choose EVs. Furthermore, governments are introducing financial incentives, such as tax cuts and tariff discounts, to encourage and manipulate the EV market, effectively subsidising EVs.

EVs still face significant challenges. Besides lack of charging infrastructure, they are still hampered by high manufacturing and serviceability costs. EVs production is still relatively small, compared to ICE vehicles, and manufacturing processes do not have their benefits of mass production and economies of scale. However, the manufacturing costs, especially batteries, are

---

[24] https://www.washingtonpost.com/climate-environment/2021/01/28/general-motors-electric/
[25] https://www.alliedmarketresearch.com/electric-vehicle-market
[26] https://www.bloomberg.com/press-releases/2021-03-30/electric-vehicles-ev-market-to-reach-3-105-4-billion-by-2027-exclusive-report-covering-pre-and-post-covid-19-market-analysis
[27] https://www.globenewswire.com/news-release/2020/11/04/2120566/0/en/Commercial-Vehicles-Market-Size-Hit-Around-US-2-55-Trillion-by-2027.html#:~:text=The%20global%20commercial%20vehicles%20market,6.8%25%20from%202020%20to%202027.

dropping rapidly.

Currently EVs constitute a small proportion of overall vehicles fleets. Despite the undeniable rapid growth in demand for EVs, which is being helped by favourable legislation, the number of ICE vehicles on the roads will continue to be dominant for the next few years. It is forecast that although EVs growth will be gradual, it will grow to be significant in the near future. We will not see a cliff edge moment, when ICE vehicles dramatically disappear. Including all their types, EVs are forecast to be only 40% of overall fleets on the world's roads by 2040 (See Figure 5.4)

There is consensus among analysts that EVs will replace ICEs eventually. However, disagreements exist regarding the time frame, with forecasts varying widely. This will be partially dependent on the development of the required infrastructure, including road services essential for EVs operations, especially making charging points and stations as easily available[28] as petrol stations are nowadays. One solution explored is to provide battery switching or "swapping stations" to cut the time of effective recharging, where a depleted battery is replaced by charged one, to reduce the time spent at charging stations to similar time at traditional petrol stations. While several companies ventured into the "charging station" space,[29] none have been hugely profitable because of the many logistical hurdles that exist, such as availability of many types of batteries at the stations and the need for experienced trained personnel to perform various service tasks. Other suggestions being considered include wireless charging[30] and road powered electric vehicles.[31]

---

[28] Although many new homes and renovated building, and some parking facilities and garages have nowadays or will require to have a charging point on the property so vehicles can be charged while idle at home, or for employees and customers in carparks, the big problem is in blocks of flats with no parking spaces. This is most of the housing in many countries.

[29] https://en.wikipedia.org/wiki/Charging_station#Battery_swapping

[30] https://www.autocar.co.uk/car-news/electric-cars/uk-firm-begins-public-trial-wireless-ev-chargers

[31] https://www.businessinsider.com/charge-electric-car-while-driving-technology-cornell-2021-5?r=US&IR=T

In the last few years, the playing field of vehicle manufacturers have transformed significantly, with companies consolidating through mergers, acquisitions and alliances. Companies are also sharing platforms and establishing multiple manufacturing bases. Many companies realised that change is needed and that they can no longer rely solely on ICE vehicles manufacturing. In order to survive, they need to join the electric car revolution. Consequently, most companies have invested heavily in developing EV models, expanding their offerings and portfolio.

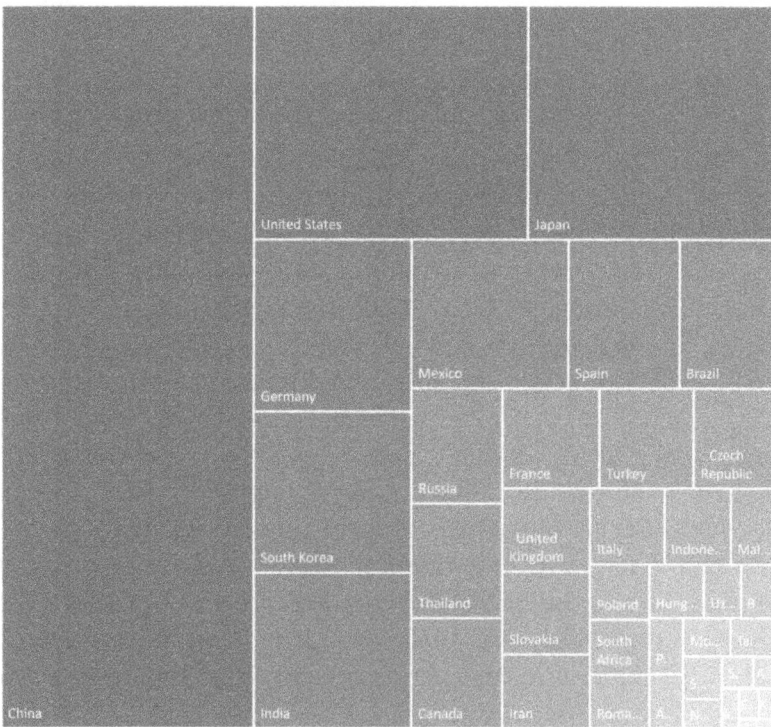

*Figure 5.5: Share of motor vehicle production by country*

Source: https://countryeconomy.com/business/motor-vehicle-production
Note 1: Numbers include passenger and commercial vehicles, but exclude two and three wheelers, and military vehicles.

At the time of writing, motor vehicle production numbers

exceeded 77 million units in 2020.[32] It can be seen that there are five top countries that dominate, accounting for almost two thirds of the total production (see Figure 5.5). The numbers above do not show the actual domination that is enjoyed by even less countries and a few big companies, since in terms of ownership, companies from these countries own production facilities in many other countries and thus, their actual dominance is even stronger.

All the big names in the industry have EV offerings - General Motors, Ford, Stellantis (Fiat Chrysler and PSA), BMW, Volkswagen, Daimler, Nissan, Toyota, Honda, Volkswagen, Renault, Hyundai, BYD, JAC Motors, SAIC, BAIC, Chery, Geely, Zotye, and Tata. However, none of them is dominant in that space. There is aggressive competition to establish a market presence and expand their market shares. They pursue different strategies, investing in different technologies, batteries and cooperate with different battery manufacturers.

Beside traditional manufacturers, several EV-only manufacturers have emerged and positioned themselves at the forefront of the automotive sector. The most important is Tesla, which at the time of writing (December 2020), has achieved a market capitalisation exceeding all the traditional manufacturers. Defying sceptics,[33] it has surpassed the combined value of the big two American automotive manufacturers,[34] or the combined big three German automotive companies. This is despite producing only 509 thousand[35] vehicles, compared to 8.9 million[36] and 9.5 million[37] by Volkswagen and Toyota, the market leaders. See Figure 5.6. Other EV- only manufacturers include the Chinese companies such as Nio (China's answer to Tesla), Xpeng Motors, JMEV, as

---

[32] https://countryeconomy.com/business/motor-vehicle-production

[33] https://oilprice.com/Energy/Energy-General/Is-The-Tesla-Bubble-About-To-Burst.html

[34] Chrysler is now part of the European company Stellantis

[35] https://ir.tesla.com/press-release/tesla-q4-2020-vehicle-production-deliveries

[36] https://www.statista.com/statistics/272050/worldwide-vehicle-production-of-volkswagen-since-2006/

[37] https://media.toyota.co.uk/wp-content/files_mf/1618251066ToyotaCompanyBackgroundv2.pdf
#:~:text=Worldwide%20production%20of%20Toyota%20Motor,2020%20was%209.528%20milli
on%20vehicles.

well as many start-ups in Europe, North America and Asia Pacific.[38]

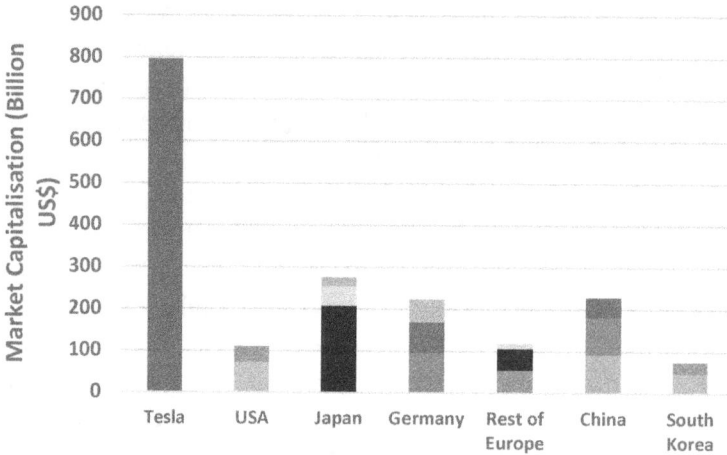

*Figure 5.6: Comparison of the market capitalisation of Tesla versus the top two or three manufacturers in major motor vehicles producing countries*

Source: https://www.visualcapitalist.com/worlds-top-car-manufacturer-by-market-cap/
Note 1: Values at the end of 2020

FCVs remain a niche product, with very few commercial models available as passenger cars, by Hyundai, Toyota and Honda. Fuel cells are being developed and tested in other kinds of vehicles including trucks, buses, boats, motorcycles and bicycles. At the moment, there is only limited hydrogen infrastructure, However the hydrogen economy is often mentioned as the new big thing. See Section 3.6.

With advances in new materials, connectivity systems, software, automation and robotics, both traditional and electric-only vehicles manufacturers are investing heavily in software and systems to automate driving functions. The ultimate aim is to achieve complete autonomous driving experience, with use of artificial intelligence (AI) a more central part. Although AI systems may define a new mobility future, nevertheless they are

---

[38] https://www.energystartups.org/top/electric-cars/

still considered to be human developments and cannot themselves be credited with inventions in patents.[39] This issue is discussed in the next section (Section 5.2.2) in more detail.

While the initial push for EVs was in passenger cars, both four seaters and two seaters, the focus is expanding to commercial medium and heavy vehicles, as well as buses,[40] which, alongside delivery vans, are projected to expand quickly as the most lucrative segments. The infamous Tesla public relations stunt, breaking its electric truck windows, brought a lot of attention to the sector.[41]

It is important to emphasize a major difference between traditional ICE vehicles manufacturing and EVs manufacturing. In traditional ICE cars, manufacturers often design and build all essential car components, in particular the engines and the bodies. Competition is fierce to design the most efficient engines, that can deliver the highest speed, while consuming the least energy. This has been the case since the introduction of the combustion engine.

In the electric cars arena, one important component is being designed, developed and manufactured separately, namely, the batteries. Battery manufacturers are often separate companies, who have supply agreements with car manufacturers.

This supply model is similar to the model implemented in the aviation industry, where engine manufacturers are also separate from aircraft manufacturers, but have co-dependency. Even though some companies, such as Tesla or BYD, have a battery manufacturing division, both companies also get supplies from other battery manufacturers to equip some of their vehicles and also supply batteries to other vehicle manufacturers.

Battery manufacturing is dominated by Asian companies, who

---

[39] https://www.bbc.co.uk/news/technology-52474250

[40] https://about.bnef.com/electric-vehicle-outlook/

[41] https://medium.com/better-marketing/did-tesla-break-that-cybertruck-window-on-purpose-b4460c100886

are building more capacity in Asia, while expanding in Europe and the US, as demand is forecast to increase more than tenfold between 2018 and 2030.[42] Currently there are 388 manufacturing locations are in Asia, 38 in the US and 23 in Europe.[43] In terms of capacity, China is dominant,[44] being home to 73%, followed by the US,[45] in second place with a paltry 12%.[46] Main batteries manufacturers are CATL, BYD and Lishen of China; LG Chem, SK Innovation and Samsung SDI of South Korea; and Panasonic of Japan. All of them supply multiple vehicle manufacturers. Tesla is the largest American manufacturer, which barely makes it to the top five in terms of production capacity, but it is expanding rapidly.

Interestingly, some vehicles manufacturers do not seem particularly eager to get involved in battery-cell manufacturing themselves. They prefer to outsource the research and development of finding the right chemistry to optimise the batteries, rather than subject themselves to sourcing of the raw materials or components, some of which have been targeted by environmentalists for dirty mining practices. In addition, by using multiple suppliers they keep their options open and can switch to better technology more easily if it is developed.

From 2010 to 2019, lithium-ion battery prices (when looking at the battery pack as a whole) fell by 87%. From 2018 to 2019 alone, the cut was 13%.[47] Tesla plans to introduce a new low-cost battery with a longer range for its Model 3 in China. This improvement will bring the cost of the car in line with gasoline vehicles.[48]

---

[42] https://seekingalpha.com/article/4289626-look-top-5-lithium-ion-battery-manufacturers-in-2019
[43] https://www.statista.com/statistics/974302/battery-manufacturing-locations-worldwide-by-region/
[44] https://oilprice.com/Energy/Energy-General/The-Secret-Behind-Chinas-Battery-Dominance.html
[45] Most US locations are operated by Asian companies.
[46] https://www.forbes.com/sites/rrapier/2019/08/04/why-china-is-dominating-lithium-ion-battery-production/#44852d3c3786
[47] https://www.greencarreports.com/news/1126308_electric-car-battery-prices-dropped-13-in-2019-will-reach-100-kwh-in-2023
[48] https://www.reuters.com/article/us-autos-tesla-batteries-exclusive/exclusive-teslas-secret-batteries-aim-to-rework-the-math-for-electric-cars-and-the-grid-idUSKBN22Q1WC

There are three materials critical to battery technology: lithium, cobalt and rare earth. The limited availability of these materials poses a significant challenge to the industry and, indeed to other electronics using industries, as their production and supply relying on only a few countries (some extremely unstable). This poses significant risk to the overall industry. See Box 7.

While EVs are certainly a threat to ICE vehicles domination, eating into their market share, they are not the only factors. In the longer term, the changing attitudes towards car ownership, commuting, working from home, driving in general and MaaS can constitute bigger threats, leading to less driving, resulting in less mobility (by choice or necessity).

Another area where rapid developments have been seen recently is the production of commercial flying cars and taxis. Prototypes have been demonstrated at many events and several players announcing marketing plans. It appears that the human dream of being able to individually fly, –albeit enhanced, is coming closer. See Section 5.3.

### Box 7: Critical Electrification Minerals

Increased battery demand will create other availability problems, replacing the fossil fuels availability problems. As mentioned earlier, the availability of lithium, cobalt and rare earth metals can become critical. The main questions that arise are whether there will lithium or other cartels, with the supply chain issues of cobalt and other rare materials.

In each metal, the overwhelming production is controlled by handful of countries. Figures 5.7 to 5.9 show that Australia, the DR Congo and China dominate production of lithium, cobalt and rare metals respectively.

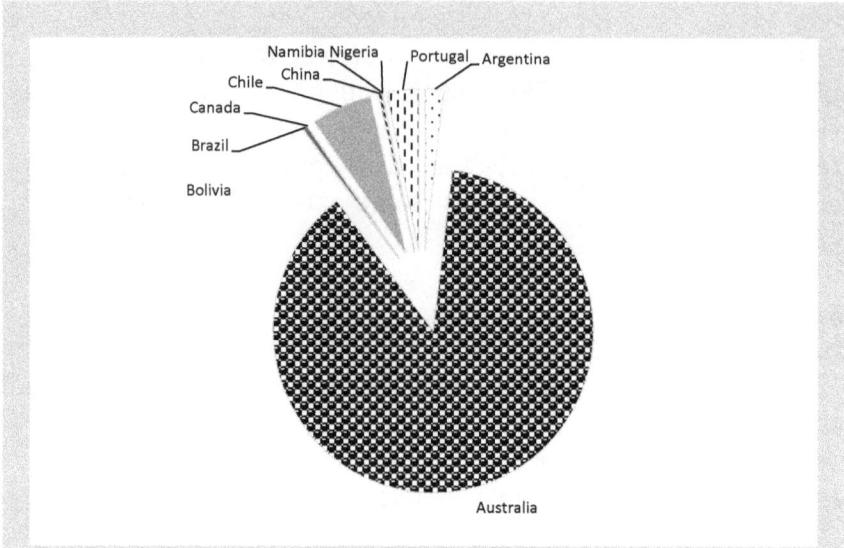

*Figure 5.7: Production of lithium minerals in 2019 (metric tonnes)*

Source: https://www.bgs.ac.uk/mineralsuk/
Note 1: Values in percentage terms.

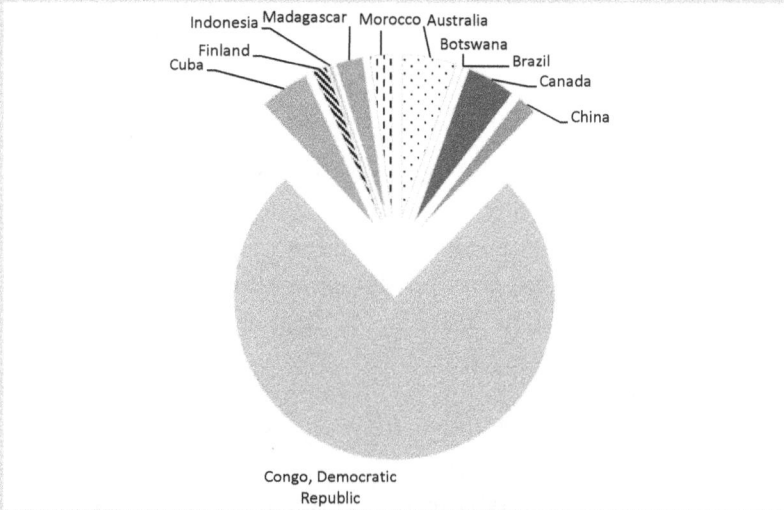

*Figure 5.8: Production of mined cobalt in 2019 (metric tonnes)*

Source: https://www.bgs.ac.uk/mineralsuk/
Note 1: Values in percentage terms.

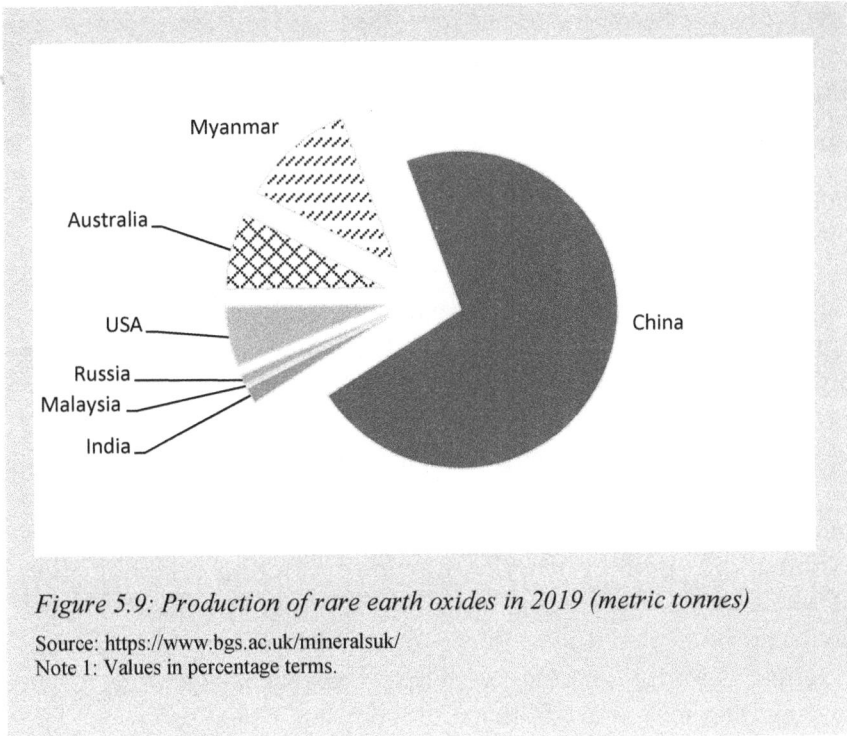

*Figure 5.9: Production of rare earth oxides in 2019 (metric tonnes)*

Source: https://www.bgs.ac.uk/mineralsuk/
Note 1: Values in percentage terms.

## 5.2.2 Transformation into Services

As discussed earlier, EVs are not the only threat to oil markets. The significant change in attitudes towards driving and car ownership could have more impact on longer term demand. The term mobility-as-a-Service (MaaS) is now universally used to describe a shift away from privately owned transportation and towards mobility provided as a service, where the aim is to offer people mobility options for their travel needs.

MaaS are often provided through gateways or platforms that connect travellers and suppliers. These solutions, nowadays mostly via apps, enable service users to book and manage trips, with payments often charged per journey or with a subscription fee, which can even be split by multiple users (ride-splitting). Some of the services can combine public and private companies' offerings. This allows travellers to organise complex travel plans involving different modes of transport and different providers conveniently transforming journeys into seamless trip chains, with bookings and payments managed collectively for all legs of the trip.[49]

Initial thinking was that these apps would be universal, allowing roaming with the same app working in different countries, without the user needing to become familiar with a new app or to sign up to new services. However, political and administrative hurdles by governments made this impossible. Many app providers were banned or had to withdraw from many markets. Consequently, the industry has fragmented into regional markets where certain apps dominate certain countries or regions.

Initially service providers offered one specific service. This has evolved with mobility service providers offering a combination of services, i.e. ridesharing, e-hailing services (where traditional taxis can be reserved via Internet or mobile phone), bike-sharing systems, scooter sharing systems, car-sharing services, van sharing services and on-demand "pop-up" bus services. The integration of many services into one platform, rather than developing one specific app per service, adds to the convenience of these platforms and has made them essential for many users. Some apps have ventured even further and are now offering food and restaurant delivery as part of their services. With this service, they act as intermediaries between customers and restaurants or independent take-out food outlets.

This expansion is attracting a diverse array of competitors, including automakers, transportation network companies and tech companies. It is characterised by a shift from businesses pursuing a single service to many, or from Shared Mobility 1.0, to Shared Mobility 2.0. This is providing opportunities for new competitors.[50]

---

[49] Currently in London, commuters may use a contactless payment bank card as an alternative to a dedicated travel card (called an Oyster card) to pay for their travel and travel seamlessly between the multiple modes, trips, and payments. Data can be utilised to optimise public transport operability and future planning, albeit individuals sacrifice part of their privacy.

[50] https://www.luxresearchinc.com/blog/shared-mobility-poised-to-disrupt-multitrillion-dollar-

Although the various delivery apps started life independently, the integration of transportation apps with delivery apps, is seen as a logical step and natural fit. The same peer-to-peer concept[51] of delivery is applied, but the transported object differs. The aim for delivery apps is to facilitate the mobility of goods and this convenience has been welcomed by the public. It also further strengthens the arguments against car ownership.

All these services started as part of the sharing economy (or gig economy). This is the new way of purchasing goods and services, that differs from the traditional business model of corporations hiring employees to produce products to sell to consumers. In the sharing economy, individuals are said to hire out things like their cars, homes and personal time to other individuals, in a peer-to-peer fashion. The apps offer platforms to facilitate that sharing and they earn their money by taking a cut from the exchanged payment. The apps are evolving into multi-service ecosystems or e-markets, where they can become one-stop-shops, providing multiple services and in performing many of these functions, substitute the use of cars and certainly make the need for car ownership unnecessary.

Some providers are exploring venturing into flying-taxis, passenger drones and similar innovative services. Many developers are testing prototypes of the vehicles anticipated to start providing these services.

Some of the MaaS apps are now considered essential parts of people's everyday life. Top of the list is Uber, it started as ridesharing company and introduced the term "uberisation" as the equivalent to ridesharing. It then expanded into micro-mobility and meals delivery. It recently announced a service offering helicopter rides in New York City[52] and is preparing to launch a flying service or "air taxi", as well. Uber operates in over 60

auto-industry
[51] A peer-to-peer (P2P) service is a decentralized platform whereby two individuals interact directly with each other, without intermediation by a third party.
[52] https://www.uber.com/gb/en/ride/uber-copter/

countries and, at the time of writing, its market capitalisation had reached US$100 billion, surpassing the combined market value of General Motors and Ford.[53]

Uber's main US competitor is Lyft, while its international counterparts include Didi Chuxing, Grab, Ola, Gett, Bolt, Gojek, Cabify, and Yandex. Some, such as Lime, Bird, and Nextbike provide micro-mobility (e.g. bike and e-scooter) services only. While others only offer integrated transport system mapping and data, such as Transit, CityMapper and Moovit.

In the delivery arena, several companies have grown to become international multi-billion US$ establishments, offering services internationally. Amongst them are Delivery Hero, GrubHub, Just Eat, Deliveroo, Meituan-Dianping, Rappi, and Zomato.

In transportation, MaaS is basically an evolutionary process. In the early 2000s, car clubs disrupted the car rental business model by providing services where people rent cars for short periods of time, often by the hour. Car Clubs differ from traditional car rental in that the owners of the cars are usually private individuals and the carsharing facilitator is a service provider, distinct from the car owner. The simplicity of accessing cars, via apps, resulted in a boom in the market. Traditional car rental companies adopted the model, launching their own carsharing services. In 2013, Avis Budget Group purchased Zipcar, the carsharing market leader, for approximately US$500 million, making Zipcar its subsidiary.

The carsharing model continues to evolve. Currently apps exist to facilitate shared rides. The next disruption will be where owners offer their cars directly to passengers without middlemen. The development of tracking apps for Covid-19 may offer an opportunity to develop the "peer-to-peer" direct contact, which is ideal for these interactions.

Furthermore, the market is also expanding to home removal services. The apps can be used as platforms matching a user's

---

[53] https://uk.finance.yahoo.com/; December 30th, 2020

delivery route to that of a transport provider's and connects them, this minimises costs and cuts emissions by optimising storage space and haulage.[54]

MaaS will eventually cause a decline in car ownership. The timing of this will differ from country to country, depending on local factors, as well as the convenience and availability of MaaS. The trend will in turn be alleviated by the advancement of self-driving cars. This raises the question of the economic benefit in privately owning a car, as opposed to using on-demand car services, whose costs will decline in the future, when cars can autonomously be self-driven.

The promise of autonomous driving (self-driving, advanced autopilot) has attracted stakeholders from different sectors. Vehicle manufacturers (e.g. General Motors, Ford, Toyota, Nissan, BMW, Mercedes-Benz, Tesla, Local Motors[55]), MaaS companies (e.g. Uber), automotive components (e.g. Local Motors – via Delphi Technologies) and technology companies (e.g. Alphabet, Apple, Intel – via its subsidiary Mobileye, Yandex, Baidu) have announced plans to develop autonomous vehicles and some are already in the testing phases. There are rumours that others are also developing autonomous vehicles too. Development is currently spearheaded by technology companies. In particular is Alphabet, Google's parent company, whose subsidiary, Waymo, is at the forefront of the development cycle and it mainly concentrates on retrofitting, or equipping existing vehicles, with self-driving capabilities. Besides this, several start-ups have gained a high profile by designing self-driving vehicles from scratch. These include Zoox, Aurora, Pony.ai and Nuro.

In the next few years, once autonomous driving becomes mainstream, travellers will be faced with a stark choice, either to prioritise convenience or privacy, as one comes at the expense of

---

[54] An example in the pan-European AnyVan app.
[55] Local Motors is an American motor vehicle manufacturing company, established in 2007 and based in Phoenix, Arizona. It focuses on low-volume manufacturing of open-source motor vehicle designs using multiple microfactories.. It develops vehicles using 3D Printing and utilizes vehicle designs provided by the online community. https://localmotors.com/

the other. Advanced autonomous driving involves continuous tracking, resulting in the further erosion of privacy, as all vehicular travel will be recorded. Also, in reality, control over where vehicles can go can be controlled by operators.[56]

Some of the products recently released in the self-driving space are electric self-driving, local commerce, delivery vehicles which are designed to carry cargo only. Nuro has used such vehicles in partnerships to test the fully autonomous delivery of groceries or pizzas.[57] Delivery by drones can form part of this evolving ecosystem and has been tested extensively in the last few years, with Amazon amongst the main stakeholders pushing for its development and adaptation. The concept gained credibility particularly in the midst of the pandemic, where trials conducted in the UK, delivering medical supplies to the Isle of Wight, gained front page coverage.[58]

Moreover, trucking in general and long-distance trucking in particular, is seen as being at the forefront of adopting and implementing the technology, with several parties developing and testing the technology. These parties come from different sectors - vehicles manufacturers (e.g. Daimler, Volvo), MaaS (e.g Uber), defence (e.g. Lockheed Martin), specialised (e.g. Starsky Robotics) and equipment manufacturing (e.g. Caterpillar, Komatsu). Companies such as the Codelco Chilean State Mining Company, Suncor Energy (a Canadian energy company), and the Rio Tinto Group, were among the first to replace human-operated trucks with driverless commercial trucks run by computers.[59]

It is clear to all that the world is changing incredibly fast. Methods of transporting people and goods from point A to point B are changing too. Increased urbanisation is changing mobility patterns and the sharing economy is changing ownership models.

---

[56] One convenience of autonomous driving will be the elimination of drinking and driving phenomena.
[57] https://en.wikipedia.org/wiki/Nuro
[58] https://uk.reuters.com/article/uk-health-coronavirus-drone/delivery-drone-flies-medical-supplies-to-britains-isle-of-wight-idUKKBN22O2PD
[59] https://en.wikipedia.org/wiki/Self-driving_truck

Automotive manufacturers in developed countries may see a substantial reduction in the number of cars they sell as car sharing becomes mainstream. The new living model of convenience is supported by successful delivery services. Mobile phone technology, which has essentially placed a supercomputer in everyone's pocket, allows endless opportunities for connectivity. This will only increase with the emergence of the internet of things, connecting trillions of things. This emerging new reality, defining mobility's future, is threatening the automotive industry and by association, the oil industry, where demand will inevitably decline.

As a result of Covid-19 pandemic, the "working from home" concept has proven itself as a viable model for the future of work. This will have major implications on commuting and the use of transport systems. The effect will not only be disruptive to car sales and ownership, but can also be disruptive to MaaS companies, reducing overall demand for ridesharing and bike sharing, if workers continue to spend more time at home. As an illustration of the potential, the market capitalisation of Zoom, the videotelephony and online chat services provider, soared in May 2020, surpassing the market capitalisation of General Motors, or the top five US airlines combined.[60] Thus even videotelephony apps can now be considered a threat to business travel and commuting.

### 5.2.3 *Arab Positioning – Glorified Camel Riding*

When I discussed the Arab world's policies and actions in relation to energy transition, my initial pessimism became cautious optimism. In many Arab countries, despite significant reliance on oil and natural gas, initiatives promoting and developing alternative energy are underway. There is definitely a momentum to develop renewables like solar energy in some countries, and to promote green technologies, such as hydrogen and electrification.

---

[60] https://uk.finance.yahoo.com/; May 17th , 2020

But, upon examining the other energy guzzling sector in the Arab world – transportation, the picture is not quite so positive. Despite being one of the major global markets for vehicles, almost all of the vehicles are imported. High end luxury car brands consider the Arab world to be one of their major markets and all have a significant retail presence there. There is no indigenous motor industry and certainly very little research and development into non-ICE vehicles. With the exception of the growing vehicles assembly facilities in Morocco and some infant projects in Egypt, and Algeria the situation in other Arab countries is not encouraging. Overall, the Arab world remains purely a vehicle importer. Arab countries gain significant revenue by imposing customs duties, tariffs and taxes on imported vehicles. Hence, despite declaring intentions to the contrary, those governments see no incentive in developing a local vehicular manufacturing industry.

Just consider that Egypt started an automotive industry almost at the same time as South Korea and China. Both China and South Korea grew to become automotive power houses, with the former becoming the world's largest producer since 2009 and nowadays produces almost 33% of the world's motor vehicles, and the latter climbing the quality ladder where its brands are now seen as reliable and even luxurious. China came from hardly any base, established JVs with western companies then progressed into its own independent industry. South Korean Hyundai grow from humble beginnings to top 5 company by 2005, and surpassed GM to third in 2015.

Growing up, I remember when the first Korean car (Pony) was imported to the Middle East it was the butt of all jokes. Was fear of failure or being ridiculed the reason Egypt gave up that opportunity?

In Egypt sadly, the industry went backwards for decades. Despite a turnaround and something of a revival in recent years, the industry has stagnated and is reduced to the role of minor assembly partners. The Egyptian assembly lines usually produce the same model for several years, unlike the competition who produce new models, in various ranges, annually. There is minimal design or innovation, and no new companies are starting out with the goal of developing their own, new, unique vehicles. Recently, Egypt reached out to China to establish assembly plants for electric cars.[61]

In the arena of electric cars, both China and South Korea are technology leaders and along with Japan, they dominate the batteries industry. The Arabs have next to zero in this domain!

In terms of MaaS, the Arab world is eagerly participating. Despite initial hostility from the governments, Arab entrepreneurs have climbed on the bandwagon of e-commerce and mobility apps, developing household name applications such as Careem, for car hiring and Talabat for food delivery. Initial hostile policies against mobility apps, i.e. Uber and Careem have mellowed, especially after Arab sovereign funds invested billions in the former. Careem development, then sale, to Uber, was a smart move. Investment in Uber may also pay off. Similarly, an investment in Lucid, a new American EV company, may have potential.

Other countries in the Middle East have enjoyed better progress in the mobility sector, with both Iran and Turkey developing their motor industry with significant volumes. While Israel, despite closing its previous motor vehicle industry, is involved in many technologies used extensively in motor vehicles - self-driving, navigation, telecommunication, also including household technology software and applications such as Mobileye, Waze, and Moovit.

Arabs are well known for their love of sporting competitions – horse racing, golf and football. They also host Formula One car races and motorbike races. While this creates huge publicity, it has not led to the development of any local motor industry or localised technology.

---

[61] https://www.electrive.com/2020/02/24/chinese-ev-manufacturing-to-take-off-in-egypt/

To conclude, it is clear that the Arab world has no active research or development within terms of mobility. All new developments have, so far, passed the Arabs by. At present, they are just pure consumers. In this domain they hardly possess any patents or technology. The "Brain Drain" means that all the brilliant Arab engineers go West and stay there. With hardly any research or development in mobility and with very few factories, the Arabs will be big losers, since the new mobility revolution will have a massive impact on fossil fuel consumption.

However, although currently at the back of the race, as the underdog and the unexpected competitor, the Arab world can rally and eventually succeed. Everyone loves an underdog! But, remember, China started its industry producing rubbish cars, now they are world leaders. If research and development is supported, some start-ups will have the advantage of avoiding the mistakes made by others and flourish by selecting successful models to improve upon or be inspired by.

## 5.3   Air Transport

Air transport is the second most substantial energy-consuming transport mode. Several main elements must combine to enable a fully functioning air transport system. i.e. aircrafts, infrastructure (airports, air navigation services, software), operators (i.e. airlines) and passengers.

Passengers' customer experience is often rated on their interaction with airports and airlines. However other elements are equally essential, without them air transport cannot operate. Amongst these are the air navigation services, which are often provided by public, or private, legal entities providing air traffic control and management, communication navigation and surveillance systems, meteorological service for air navigation, as well as search and rescue services.

Air transport fortunes are linked tightly to oil price. Fuel costs constitute a significant proportion of the sector's operating costs and can determine the prosperity and survival of the air transport sector. The two elements affected directly by oil price are the airlines whose operations consume fuel and aircraft manufacturers, who strive to increase fuel efficiency of their aircraft to reduce fuel consumption. Since fuel cost is the most important expenditure for airlines, manufacturers compete aggressively to produce more efficient aircraft, investing heavily in research and development to optimise fuel consumption. At the same time, oil producers have a vested interest in supporting a robust air transport sector, to boost oil demand and sell their refined products. Any significant reduction in air transport impacts badly on the fuel sellers' finances and consequently on oil refiners and producers.

In the remaining section of this chapter, I briefly examine the subject of air transport, focusing on aircraft manufacturing companies and airlines, as they are the two most significant drivers interacting with fossil fuel markets. I have limited my analysis to civil air transport, as military air activities are outside the scope of this book.

The terms aircraft and aeroplane[62] are often confused. Aircraft is a term that refers to anything that can fly. Thus, this term includes aeroplanes, helicopters, airships, gliders, powered paragliding (paramotors), ultra-light aircraft, drones, unmanned aircraft, robotic aircrafts and hot air balloons.

Aircrafts are classified by different criteria, such as:

- Lift type: lighter than air – aerostats, including hot air balloons and airships; or heavier-than-air – aerodynes, which are divided into two main sub-classes fixed wing and rotorcraft. Combinations of the two subclasses and other lift methods are marginal
- Propulsion: unpowered aircrafts; powered aircrafts, which

---

[62] Airplane and aeroplane are the same thing, but aeroplane is the preferred British spelling.

are divided into propeller aircrafts jet aircrafts, rotorcrafts, and rocket engines.

- Size; ranging from toy-size or drone size to tens of metres of length or wingspan and a few hundred tons weight.[63]
- Speed, which can reach hypersonic.
- Range; i.e. the distance an aircraft can fly between take-off and landing.
- Usage; military; civilian, experimental, and model.[64]

The classifications overlap so a certain aircraft can fall under many classifications, depending on the criterion it is viewed with. Thus, an aeroplane is a powered, fixed-wing aircraft, with a propeller, a jet engine or a rocket engine.

Civilian aviation is categorised as commercial or general aviation. Both categories involve the transport of passengers, freight/cargo and mail.

Commercial aviation involves operating aircraft for hire, including scheduled and non-scheduled flights (charter, and on demand, air-taxi, commercial business aviation).

General aviation,[65] on the other hand, includes all civil aviation operations, other than scheduled air services and non-scheduled air transport operations, for remuneration or hire. Besides private and recreational travel, it often encompasses aerial work[66] used for specialized services, i.e. agriculture, construction, photography, surveying, observation and patrol, search and rescue, aerial development, etc. This sector operates all types of aircrafts including fixed-wing, rotorcraft, gliders and balloons. Note that business aviation can be either commercial or general, depending on whether there is a charge for the service provided.

---

[63] https://edition.cnn.com/travel/article/worlds-largest-airplanes/index.html
[64] A model aircraft is a small unmanned type made to fly for fun, for static display, for aerodynamic research or for other purposes.
[65] https://www.skybrary.aero/index.php/General_Aviation_(GA)
[66] Although the International Civil Aviation Organization (ICAO) excludes any form of remunerated aviation from its definition, many countries include aerial work as part of general aviation.

The aviation market is dominated by global aviation, in terms of aircraft numbers, number of airports used and traffic volume. However, in terms of number of travellers, revenue and fuel consumption - commercial aviation dominates. As of 2019, there were approximately 27,500 commercial aviation aircraft, with 29% in North America.[67,68] In comparison, in 2019 there were more than 440,000 general aviation aircraft in the worldwide fleet, ranging from small training aircraft and helicopters to intercontinental business jets. About 211,000 of these general aviation aircraft, or 48%, are based in the US. There are more than 5,100 public airports in the US with fewer than 400 airports served by commercial airlines.[69]

To put the market size, in terms of aircraft sales, in context, in 2019, the commercial aircraft market, was valued at just over US$191 billion,[70] whereas the general aviation market was valued at US$20.6 billion, excluding helicopters.[71,72] The global commercial helicopter market size was valued at US$5.3 billion.[73]

The numbers above illustrate the dominance of the commercial aeroplane sector, often referred to as "airliners". Airliners are often classified into four classes, by size and weight and, when they are used to carry passengers, the classification often correlates these classes to the number of passengers an aeroplane can carry. These classes, from largest to smallest, are:

- Wide-body airliners, often called twin-aisle because they generally have two separate aisles running from the front to the back of the passenger cabin. These are usually used

---

[67] https://www.oliverwyman.com/content/dam/oliver-wyman/v2/publications/2019/January/global-fleet-mro-market-forecast-commentary-2019-2029.pdf
[68] The number increases to roughly 60,000 aircraft including cargo and charter, https://www.iaopa.eu/what-is-general-aviation
[69] http://www.fi-aeroweb.com/General-Aviation.html
[70] https://www.researchandmarkets.com/reports/4583619/the-global-commercial-aircraft-market-2018-2028
[71] http://www.fi-aeroweb.com/General-Aviation.html
[72] https://gama.aero/wp-content/uploads/GAMA_2019Databook_Final-2020-03-20.pdf
[73] https://www.grandviewresearch.com/industry-analysis/commercial-helicopters-market

for long-haul flights between airline hubs and major cities.

- Narrow-body, often called single-aisle. These are generally used for short to medium-distance flights, with fewer passengers than their wide-body counterparts.
- Regional airliners, which typically seat fewer than 100 passengers and may be powered by turbofans or turbo-propellers. These are used to feed traffic into the large airline hubs and provide domestic regional air transport.
- Light aircraft, often called commuter aircraft, commuterliners, feederliners and air taxis, depending on their size, engines, seating configurations, location of operation or marketing strategies. They are propeller powered aircrafts, single or double-engine. These are used as short-haul regional feeders and carry a small number of passengers, 19 or fewer.

Although the airline industry overall is an extremely competitive market, there is little competition among its suppliers of aircraft. When we look at aeroplanes manufacturing, we can see that, compared to road transport, considerably fewer manufacturers exist in this sphere. Note that many of these manufacturers have dual roles, where they operate in both the military and the civil aviation markets (e.g. Boeing and Airbus), whereas others generate most of their revenue from military products (e.g. Lockheed Martin, and BAE Systems).

The wide-body and narrow-body airliner market is limited to a handful of manufacturers in US, EU/UK, Russia, China, and Ukraine. Airbus and Boeing dominate the market, enjoying an effective duopoly.

The regional airliner markets have more manufactures, in addition to the big manufacturers, with even more players in the EU, Brazil, Canada, and Japan involved. The big four in regional airliner markets are ATR Aircraft (Aerei da Trasporto Regionale or Avions de transport régional), Bombardier Aerospace, De Havilland and Embraer.

Finally, the light aircraft market is more diverse, with a large number of manufacturers worldwide. The top five, by sales revenue[74] include Gulfstream (a subsidiary of US defence and aerospace giant General Dynamics, the largest manufacturer by sales values), and Textron Aviation (owner of Cessna and Beechcraft) of the US, Dessault in the EU and two dominant regional aircraft players, Bombardier and Embraer. Other notable names include Piper, Cirrus, Pilatus, Daher, Tecnam, Diamond, Air Tractors and Honda.

Helicopter manufacturing market is diverse, with companies operating in over 30 countries, many of them producing both civil and military vehicles. The major companies are Airbus, Leonardo, Bell Textron, Robinson, Sikorsky (a division of Lockheed Martin) and Russian Helicopters.

Helicopters are important in the oil and natural gas sector. They are essential to provide transport and logistical support for the offshore oil and gas operations.

It has to be noted that although most planes land on land, there in a niche market for seaplanes, amphibious planes and aircrafts. However, they comprise a tiny proportion of the market.

Aircraft manufacturing comprises a system of different components, where a vast network of companies specialises and supply certain components. Because of the extensive range of skills and facilities required, no single company builds an entire aircraft. Engines are often manufactured by several different providers. The same is also applicable to the telecoms and connectivity systems that are provided by others.

The companies discussed above are the airframe manufacturers and are always recognised by the public as the aircraft builders who are responsible for overall certification. Although the airframe manufacturers remain the major integrators and sellers

---

[74] http://www.fi-aeroweb.com/General-Aviation.html

of aircraft, substantial costs of production have shifted toward the key secondary systems of propulsion (i.e. engines) and avionics (electronic systems used in aviation) and auxiliary equipment (e.g. landing gear or armament, in the case of military aeroplanes).

The major propulsion manufacturers, who produce the full range of aerospace engines are: GE Aviation (a division of General Electric) and Pratt & Whitney (subsidiary of Raytheon) in the US; Rolls-Royce in the UK; Safran in France; United Engine Corporation in Russia and Aero Engine Corporation in China. These companies also cooperate in joint ventures such as CFM International (between GE Aircraft and Safran), and Engine Alliance (between GE and Pratt & Whitney). They also have another joint venture - International Aero Engines (between Pratt & Whitney, the Aero Engine Corporation of Japan, and MTU Aero Engines of Germany). Other manufacturers in the US, EU, UK, Canada, Russia China, Japan, India and North Korea, produce small and medium size turbofan engines. Other types of smaller engines are produced also in other countries.

Major avionics providers include Collins Aerospace, Honeywell, Northrop Grumman and Raytheon in the US, Thales Avionics in France, Panasonic Avionics Corporation in Japan, and Avionika in Russia.

With civil aircrafts, typically the costs average at 50% for structure and integration, 20% for engines, and 30% for avionics. For military aircraft, the cost of avionics, including systems associated with self-protection and weapons management, can reach 50%, with 20% for engines and 30% for airframe and integration. In fact, the classic final assembly and test phases represent a mere 7–10% of the cost of modern fighter aircraft.[75] This cost breakout is being optimised by rapid adopting of technologies.

Throughout the aircraft manufacturing ecosystem, suppliers are

---

[75] Encyclopaedia Britannica

employing new technologies to optimise their operations and costs. They are using machine-learning techniques, such as artificial intelligence (AI), to improve safety, productivity and the overall quality of aircrafts. Machine learning algorithms collect data from machine-to-machine and machine-to-human interfaces, then use advanced data analytics to drive effective decision making.[76] In addition, the impact of 3D printing technologies is making significant improvements in the manufacturing of lightweight aircraft parts and the development of smart glasses in the aircraft industry.

Market research reports in late 2019 were unanimous in forecasting that the global commercial aircraft manufacturing industry would expand at a faster pace over the next five years, anticipating continued global economic growth and the industry in emerging markets also continuing to grow.[77] Prior to Covid-19 outbreak, air transport was expanding rapidly, with growing demand for passenger traffic, which increased by 68% since 2010,[78] and was expected to grow by another 4% in 2020.[79] Similarly since 2010,[80] freight/cargo volume grew by 50% with another 2% growth forecast for 2020.[81] Pre-Covid-19 pandemic, air transport was by far the fastest growing mode of public transport. It remains the most efficient, in terms of reducing journey times, while also providing convenient facilities and reasonable comfort level for the passengers.

However, the Covid-19 pandemic has changed all those projections and has the new health guidelines have become a major challenge to both public transport and air travel, as social distancing cannot be achieved without a drastic reduction in air travel capacity.

---

[76] https://www.thebusinessresearchcompany.com/report/aircraft-manufacturing-global-market-briefing-2018

[77] https://www.ibisworld.com/global/market-research-reports/global-commercial-aircraft-manufacturing-industry/

[78] https://www.statista.com/statistics/564717/airline-industry-passenger-traffic-globally/

[79] https://www.iata.org/en/pressroom/pr/2019-12-11-01/

[80] https://www.statista.com/statistics/564668/worldwide-air-cargo-traffic/

[81] https://www.iata.org/en/pressroom/pr/2019-12-11-01/

As the companies that provide air transport services for travellers and freight/cargo, airlines are the entities that interact most with the public. They often bear the brunt of crises in air transport, as we have seen during the Covid-19 pandemic and they usually face the wrath of the public for the failings of the air travel industry.

Interestingly, even though airlines generate massive revenues, their margins are relatively low and their track record in profitability is poor. The last two decades have been extremely challenging where many airlines have posted huge losses, been forced to ask for bailouts, suffered from episodes of bankruptcy or simply gone out of business. This is not surprising once we realise that every member of the air transport chain is far more profitable than the airlines. This has been demonstrated in several studies conducted over the last three decades covering periods 1992-1996,[82] 2000-2013, and 2013-2018.

These studies also identified the cyclical nature of the airlines business model, with bouts of boom and bust that are often connected to oil price volatility. They are also negatively affected by major global events that erode confidence in air transport, such as the global financial crisis of 2008, then the aftermath of Al-Qaida attack on the US on September 11th, 2001.

Fuel costs are the biggest single expenditure incurred by airlines, which explains the close relationship between oil price and the airline sectors health and profitability. Thus, at times when oil price drops, airlines tend to thrive and when oil price soars, airlines suffer.

Fuel efficiency has improved steadily over the last twenty years, reducing the fuel consumption and thus the fuel bills for the airlines. Continued fuel efficiency gains averaged more than 2% yearly since 2009.[83] Thus in the last decade, periods when fuel

---

[82] Fundamentals of Air Transport Management Kindle Edition, y P S Senguttuvan, Excel Books; First edition (23 May 2012)
[83] https://www.iata.org/en/iata-repository/publications/economic-reports/airline-industry-economic-performance---december-2019---report/

costs were high made it economic to retire older aeroplanes and replace them with higher fuel efficiency aeroplanes. In a World Bank report in 2012, energy efficiency in the air transport sector, achieved from technology improvements in airframe and engine design, air traffic control and airport operation, resulted in the stock of aircraft being 80% more fuel efficient than their 1960s counterparts.[84] Improvements have accelerated since then and in 2019, around half of aeroplanes deliveries replaced existing fleet, making a significant contribution to increasing fleet fuel efficiency.[85]

The interconnectivity between oil price and airlines prosperity was broken in 2020. The double whammy of the oil price crash and travel restrictions to control Covid-19 virus spreading, prevented the airlines benefitting from the low oil price, which ordinarily would have boosted their traffic. On the contrary, the closure of borders, airports, etc. brought air transport to a virtual halt.

The impact of the pandemic devastated airlines, with more than 90% of their traffic cut. Many asked for state aid or bailouts. Many are on the verge of collapse. While airline finances are outside the scope of this book, the effects of reduced travel directly influence oil demand and oil markets.

In a matter of a few months, the market capitalisation of major players in aircraft manufacturing and airlines collapsed. Figure 5.10 illustrates the severity of the capital destruction. In numerical terms, the companies in the representative sample lost over US$400 billion between January and May 2020. This will lead to a new wave of consolidation and more integration between defence, aerospace and IT industries. Collapse of businesses in aviation in 2020 was unprecedented. Recovery, in whatever form it appears, will take years.

---

[84] Daniel Martinez, Ben Ebenhack and Travis Wagner Energy Efficiency - Concepts and Calculations, 1st Edition, Elsevier Science, 2019
[85] ibid.

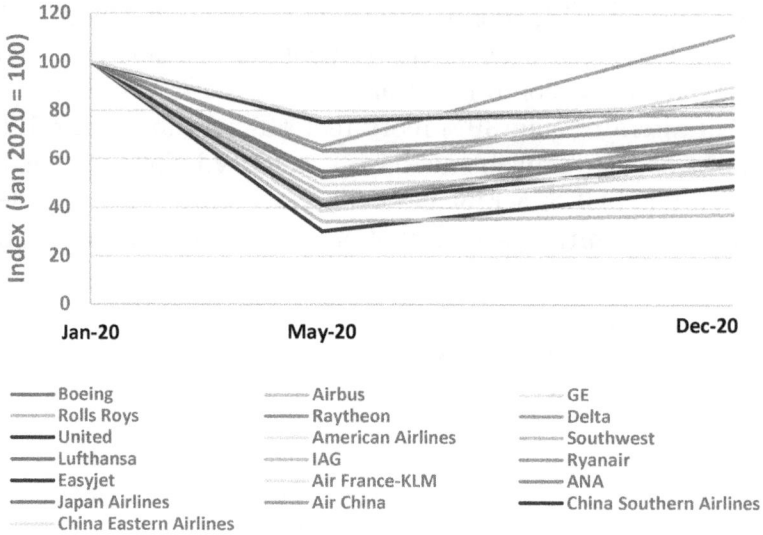

*Figure 5.10: Aviation companies share price relative to January 2020 (Representative sample)*

Source: Google Finance

In the last few years, we witnessed rapid advances in "unconventional" air transport concepts and modes, manned and unmanned - gliders, drones, satellites, rockets and space travel. Some of these devices are gaining considerable traction. Furthermore, advances in personal domestic flying transport, such as flying taxis, are blurring the boundaries between mass commercial air travel and more customised modes.[86,87] The developers of these new trends are often a cocktail of companies with diverse activities including civil aviation, defence, telecommunication, IT and even space research. An example of the latter's involvement, is the plan of high-altitude hypersonic travel, cutting travel time from Sydney to London to 90 minutes in a rocket like plane.[88] Or the expanding commercial space

---

[86] https://en.wikipedia.org/wiki/Air_taxi - see list of companies

[87] https://en.wikipedia.org/wiki/Volocopter

[88] https://www.traveller.com.au/hypersonic-jet-could-slash-flight-time-from-sydney-to-london-to-

travel spearheaded by SpaceX, Blue Origin and Virgin Galactic – each supported by a visionary multibillionaire.[89]

For years electric aviation has been extensively researched, particularly flying electric aircraft, including manned and unmanned aerial vehicles.[90] Several companies are engaged in electric aviation, including Airbus and Eviation, with over 170 projects underway globally. Electric aviation could reduce fuel costs by over 90% and maintenance cost by over 50%. However, electric planes will be limited by the distance they can fly.[91]

### 5.3.1    Arab positioning – Remembering Abbas bin Firnas[92]

In Arab circles, those interested in aviation, fondly tell  of the historic figure Abbas bin Firnas (809 - 887 A.D.) who, it is said, "flew faster than the phoenix in his flight when he dressed his body in the feathers of a vulture".[93] Caught up in nostalgia and forgetting reality, still they dream of winged men and flying carpets.

Now things are very different, when Arab air transport is mentioned, the world has a very positive perception. Emirates, Etihad and Qatar Airways are household names, flying people around the world, known for exceptional customer service, generous sponsors of sporting events and different sports teams.

In this domain, Arab countries have pulled out all the stops in developing civil aviation services, with the Gulf countries

---

four-hours-say-uk-space-agency-h1icaa
[89] https://www.revfine.com/space-tourism/
[90] For an overview, see: https://en.wikipedia.org/wiki/Electric_vehicle
[91] Scientific American, 323, no. 6, Dec 2029, p. 34
[92] Abu al-Qasim Abbas ibn Firnas ibn Wirdas al-Takurini (c. 809/810 – 887 A.D.), also known as Abbas ibn Firnas (Arabic: عباس بن فرناس), latinized Armen Firman (not confirmed), was an Andalusian polymath: an inventor, astronomer, physician, chemist, engineer, Andalusian musician and Arabic-language poet. He was reported to have experimented with a form of flight. According to John Harding, Ibn Firnas' glider was the first attempt at heavier-than-air flight in aviation history. Reference: https://en.wikipedia.org/wiki/Abbas_ibn_Firnas
[93] Lynn Townsend White, Jr. (Spring, 1961). "Eilmer of Malmesbury, an Eleventh Century Aviator: A Case Study of Technological Innovation, Its Context and Tradition", Technology and Culture 2 (2), p. 97-111

competing with each other to create luxurious top-class airlines. The top three airlines (Emirates, Etihad, and Qatar Airways) have excelled, surpassing their Asian, American and European counterparts.

While this strategy has led to enviable reputations, it unfortunately did not make them profitable. Many of these airlines, despite high revenues, would be loss-making companies, if they were not receiving favourable tax breaks and the patronage of government owners, who provide cash for expansion, which gives them unfair leverage in business.

State of the art airports have been built as homes for these airlines in the region, creating very significant international hubs. Dubai tops the list, followed by Doha and Abu Dhabi.

Despite these incredible infrastructures, those organisations depend fully on Western technology, Western machines and Western management. They have indeed invested vast sums of money, bankrolling the manufacturers but, with the exception of some localised maintenance services, they made no efforts to establish any local industry.

Consequently, despite their international success and their prestige, I am sceptical, as much of this is carefully presented image, articulately planned and run, mostly by expatriates. While the Arab host countries have benefited, improving their images and while the airlines have contributed to significant growth in other sectors, especially tourism, the impact of developing any localised aviation technology is minimal.

Why is this? Why concentrate on airlines? Is it a deliberate choice or is it bad planning? What is the wisdom in investing in the lowest margin part of the air travel supply chain? Maybe it is a passive embracing of increased tourism and all that entails, which is not insignificant? But that is surely limited and short-sighted. Maybe because it is more visible? Perhaps it is to do with ego and of the prestige of the organisations? Or is it because there is great opportunity for corruption, spending money with little traceability and without accountability? These questions are thought provoking and the answers can be speculative, so I leave it to the readers to make up their own minds.

The truth is that there is no doubt that having a successful airline has economic benefits encouraging tourism, boosting trade, creating jobs, creating economic hubs at airports, cultural mix, etc. But equally important is investment in the more profitable components of the air transport supply chain and reducing the dependence on external providers, businesses and countries. It is both ironic and disappointing that this vision is lacking in the Arab world.

Moreover, there is a catch-22 in the relationship between the airlines and their petrostates owners. For airlines, lower fuel costs improve profitability, but this, in turn, reduces revenues and weakens the economies of the Arab petrostates, who are the airlines' sponsors and patrons and who prop up their airlines financially.

In the service sector, there are promising activities in Arab world. These include the flying taxi service in Dubai, and promoting future Space travel, with establishment of Ras Al Khaimah airport, Morocco also has a landing site as emergency for space shuttles.

Similar to the situation in road transport technologies, the Arab world has no active research or development to compete in aviation technologies. Here also in aviation, all new developments have passed the Arab world by. The Arab world remains a taker of ideas, an importer of machinery and technology, a pure consumer. The Arab countries possess hardly any patents or technology, with only a handful of factories engaged in basic helicopter assembly and other basic machinery. If the situation does not change and improve, the Arabs will be at a significant disadvantage. Back in the 1950s, they were equals to China. Alas! Modern China has established a promising aviation industry, designed and manufactured its own fleet of planes. As it is the biggest buyer of aircrafts, it intends to buy local thus breaking the duopoly of Boeing and Airbus.

Other countries in the Middle East have enjoyed better progress in the aviation sector. Iran and Turkey are developing drones and unmanned aeroplanes. Israel is involved in many technologies used extensively in aviation – particularly navigation and telecommunication. Israel has been making advances in the drone industry and experimented with "delivery by drones" in Iceland since 2018,[94] and also used drones to deliver Covid-19 medicines.

---

[94] https://www.calcalistech.com/ctech/articles/0,7340,L-3743969,00.html

## Chapter 6
# *THE FUTURE IS AMBER*

If "a week is a long time in politics", five years is definitely a long time in the world of technological advancement. Much has changed since I wrote my last book. However, compared to previous historical periods, I note that the changes in this latest five-year period are unprecedented, particularly the events of the year 2020, which have been transformational, following the Covid-19 pandemic, its impacts on human life and the energy markets.

The energy transition has sharply accelerated, with several concepts, such as working from home and remote operations, tested in real time and judged as mostly successful. They will become part of a "new norm" that will prevail once the pandemic subsides.

During the ten years that passed since I began writing this book series, the world has moved on. Advances in technology have surpassed all expectations and the sense that globally responsible citizens who care about the world's wellbeing and the environment, have moved from being a utopian, extremist to a popular mainstream position. With this transformation the views about energy have evolved and, for most people, oil has become a dirty word. As a result, many companies are aligning with the zeitgeist and visibly, swiftly moving away from fossil fuels to enhance their public image.

Although this unflattering image can be justified when we analyse the historic behaviour of many energy companies, the world will not be rid of oil and natural gas soon. Oil and natural gas companies by necessity, will evolve and continue to be influential in the era of energy transition. While most of these companies are publicising their green plans, they continue to invest in fossil fuels, so it is worth remembering, "Actions speak louder than words"!

Ten years ago, when concluding the first book of the series *"Fossil Fuels in the Arab World: Facts and Fiction"*, several questions were posed to stimulate the thinking about "what's next" for the Arab world and how do fossil fuels (which shaped its past) continue to shape its present and future? My ultimate aim was to answer the bigger question in that book, which was "Who benefited from concealing the facts and promoting the myths and the fiction?" about the "Arab world".

At the time I hinted at what the answer might be. It was a one-word answer: "Politics".

Fast forward five years when, concluding the second book of the series *"Fossil Fuels in the Arab World: Seasons Reversed"*, I found myself faced again with the ultimate question– Who was really the ultimate beneficiary of the failed Arab Spring? And who benefited from the world's predicament in oil, climate change, industrialised poverty and human suffering?

The answer to those questions was not so simple. The true beneficiaries were diverse and are not necessarily all working together. They were shown to form a network of wealthy politicians, businessmen/women, speculators, market traders, the arms industry, weapon traders, opportunists, etc. Most shared one common motive, "Greed"!

Reaching the third book of this series *"Fossil Fuels in the Arab World: Missed the Boat?"*, the question posed is explicitly included in the title. With energy transition becoming the 'next big thing', the pressing question is therefore whether the Arab world has missed its chance to benefit from its oil and natural gas geological gift and wealth? Or has this gift turned into a curse? Will the Arab world end up with worthless resources of oil and natural gas, or do the Arab states still have a chance of maximising their assets before it is too late?

The previous chapters of this book (thank you for reading them before reaching here) provided detailed answers. These answers were complex, constructed, based on facts and a combination of various theories and arguments, to supply all the jigsaw pieces we require to solve the puzzle and see the big picture.

Completing this book was really challenging. I started the manuscript just before Covid-19 virus broke out and started systemically reviewing and researching the relevant events affecting the Arab world and the energy markets. Then when Covid-19 hit the world, all these events took second fiddle to Covid-19 pandemic, which cemented its position as the fundamental event that took place and not only dominate our lives since then, but its aftermaths will be felt for years to come. The intended book I began was utterly changed by the arrival of the pandemic.

Thus, as discussed in Chapter 1, in addition to the pandemic and often as a ripple effect of it, other major incidents took place, both globally and within the Arab world. In this short period, the Arab world changed, but in most places the direction of change was unfortunately retrograde. Many of its countries were totally devastated and the established political landscape in the region was shattered. Democratic green shoots that sprouted in 2011 were killed off by 2015 and a second wave of protest that recommenced in 2019 was ended by the pandemic. Overall, the geopolitical picture in the region has changed irreversibly.

As if things were not bad enough, the second oil price collapse in six years happened. It had significant effects on many, especially in the Arab world, where most of its countries depend on oil and natural gas revenues, either directly or indirectly (by economic aid, remittances, trade ties with petrostates).

Furthermore, the way we all communicate and get our information and news has changed completely, depriving governments everywhere of their monopoly on providing news.

Instead, where news, information or communication were concerned, social media and the internet dominated. However, many of those governments seized the moment presented by the pandemic to regain their authority. Behind the guise of emergency public health interventions, the pandemic provided previously unimaginable opportunities for tracking and monitoring citizens once again. In the opinion of many experts, and conspiracy theorists the world over, mandatory vaccinations and use of Covid-19 apps, is just a way to start a new era of extreme surveillance.

This authoritarian reclamation of power was predicted by some, e.g. Jamie Bartlett, author of "The People versus Tech", he argued that the spread of social media and digital technology will undermine democracy. He conjectures that there is a compatibility problem between democracy and technology. Institutions and regulations - like a free press, an informed citizenry, election advertising rules, etc, keep democracy working. All voters get access to the same facts and messages. But big data analysis has enabled politicians to target voters individually, with highly targeted messaging that regulators cannot easily see, exploiting people's psychological vulnerabilities.[1] This is evident in US elections and in the UK prior to Brexit.

Several signposts I predicted at the end of my second book in 2015 were examined here in Chapter 2. Most were vindicated. We witnessed the tense political situation develop in Saudi Arabia, the deterioration in the relationship between the USA, Russia and China, as well as the continuous oil price volatility.

The oil price slump has fundamentally changed the oil and natural gas markets. The second price collapse in six years, with the subsequent oil price recovery, re-established OPEC and its allies under the umbrella of OPEC+, as the force that manipulates and controls oil price.

---

[1] http://www.bbc.co.uk/news/uk-politics-43793546

The pre-existing narrative has changed completely, from peak oil supply to peak oil demand and now, how to navigate energy transition, moving away from fossil fuels, towards alternative energy. This is due to a variety of factors amongst which include combating climate change, policy pressures as a result of implementing obligations of the Paris agreement, environmental policies, the demand for energy efficiency, the shift in transport focus and various technical developments.

It is becoming evident that the world has an abundance of oil and natural gas. The market fundamentals have advanced and the development in extracting tight/shale oil and natural gas, has changed the market dynamics for ever. However, with oil price volatility reoccurring and the oil price expected to be stuck in a narrow band, oil and natural gas companies, as well as the producing countries, need to adapt and diversify their revenue sources. They need to accept that some of their reserves will be reclassified as resources and most probably remain in the ground.

In the last few years, an increasing number of basins are being labelled "super basins" by some researchers.[2] These basins (both conventional and unconventional) are capable of achieving dramatic increases in production if the economic and ESG conditions are right. These basins contain significant amounts of recoverable reserves and resources. The potential uplift from two dozen such basins will offset the decline in conventional oil production. However, there are serious doubts that this volume will ever be needed with the rapid advancement of alternative energy rendering some of it not needed.

As discussed in Chapter 3, it is clear that political economic power is shifting away from petrostates to electrostates and chip-states. Electricification is forging ahead and, with AI and machine learning rapid advancement, the future will be determined by algorithms. Many now argue that data is the new oil. This statement is not a prediction, rather it is acknowledging

---

[2]  https://www.forbes.com/sites/uhenergy/2019/02/21/its-not-just-the-permian-super-basins-are-a-global-phenomenon/?sh=378f50ac4f21

the new reality. Just compare the current market capitalisation of IT companies as opposed to oil and natural gas companies. The massive divergence should confirm this to any doubters.

## 6.1   Where will the Arab World fit in this New World?

So, where will the Arab world fit in this New World? The answer to this question is complicated. I see that it is ambiguous, "neither fish nor fowl" and that it will be perceived as neither one thing nor the other. Neither a green energy producer nor a technological innovator, neither a prosperous oil industrial leader nor a wise utiliser of all of its resources (including its people). The development of Arab countries is diverging, with the gap between the Gulf states and the rest gradually widening. Despite all the talk about energy transition, oil and natural gas are not disappearing from our daily lives in the next twenty years, probably not even by 2050. Even if, as expected, the Arab world's share in the overall primary energy mix shrinks, most of the resources that will become stranded will be from the more expensive producing countries who will no longer be able to compete. As we have seen in Section 4.1, production from the Arab world is the cheapest and thus the oil and natural gas industry will certainly continue to prosper.

As we have seen in Table 4.1, the share of oil and natural gas reserves in the Arab world, continues to shrink as more unconventional resources come into play and are reclassified as conventional. We have demonstrated that the idea that the industrial world is dependent on the Arab world to supply their energy needs, is a complete myth. The growth of resources worldwide has been substantial, as we achieve higher recovery rates along with more tight/shale oil and natural gas becoming economical to produce. Although the oil price collapse has put the brakes on increasing production from the unconventional basins for now, the technology is now well established, and production can resume substantially if the price dynamics shift.

At the moment, the Arab world still has some influence on oil markets. Other oil producers rely on the heavyweight oil producing countries, i.e. Saudi Arabia, Kuwait, UAE and Russia, to balance the market by cutting production. However, this influence is waning as the question has moved from oil scarcity to the practice of withholding oil supply in order to raise the price, thus making oil production economic for all producers. That is very different from what we were being brainwashed with over the last decade and at last, the scaremongering about oil peak threat has disappeared.

The Arab world has also significant potential in solar energy. But to succeed, it needs wise leadership that understands the potential and has the vision to develop an alternative energy narrative and be leaders in a new energy economy. While several Arab countries have talented individuals, who understand what is needed, their voices are often drowned out by the traditionalists who prefer to maintain the status quo. What is needed now are those capable of developing pragmatic practical visions for the future rather than those who are obsessed with promoting legacy projects, their mega projects, e.g. new capital cities, new records for the tallest buildings and entering the Guinness Book of Records. We do not need our 15 minutes of fame leaders, we require creative innovators who have the best interests of their fellow citizens at heart, now and in the future. We need our own Moses, to lead us out of this desert.

As electrification is rapidly occurring, its largest impact will be on the transportation sector with advances in the mobility domain, both in terms of vehicles and services. There is consensus that the transformation is underway, but there is fierce debate on the timing and when will the inflection point occur.

The mobility revolution and its implication on the population wellbeing and the energy markets, is huge, as it will have major implication on oil demand. But in this domain the Arab world is a mere bystander and consumer. Unless something changes dramatically very soon, I cannot see this as an area of development or evolution here unfortunately.

But with lockdown and more people working from home, an unexpected side-effect of the pandemic was that the restrictions worldwide might result in a baby boom! Thankfully these fears were unfounded,[3] (despite increases in condoms sales).[4]

This leads to another fundamental issue, overpopulation and its implications for increased energy demand. Even if energy efficiency improves and the consumption per capita decreases, the overall consumption may still increase, if the rate of population growth exceeds the decrease in consumption, per capita. Population growth and/or population control is a taboo subject in the Arab world and in many other conservative societies. Those seeking to address issues of overpopulation are seen as threats by many elements in society.[5] Many are fearful of the consequences of publicly broaching this provocative subject, yet while it remains unspoken, it continues to be one of the most significant threats to the continued existence of life as we know it.

In summary, although in terms of supplying future energy the Arab world appears to have a future, not only being the last feasible oil and natural gas producer, but also with its massive solar energy potential and opportunity to be a major player in the hydrogen economy, they face competition from electrostates and chip-states to define and control the future energy supply and new energy platforms in the energy transition era.

There are two major issues that potentially threaten a bright future for the Arab world – a lack of water and a significant increase in the human population. Already the water supply is reaching critical levels and the tensions between Egypt, Sudan and Ethiopia clearly indicate how acute the situation is. The latter

---

[3] https://www.bbc.co.uk/news/business-52490023
[4] https://www.bbc.co.uk/news/business-54613475
[5] It took 123 years to reach 2 billion, and only 33 years to reach 3 billion. The last several billion milestones (4, 5, 6, and 7 billion) have been reached in 14, 13, 12, and 12 years, respectively. We became a world of 7 billion in 2011. If this trend continues the human race threatens to destroy its habitat and, in so doing, perhaps create extinction for itself.
Quoted from (https://populationconnection.org/world-population-milestones-throughout-history)

issue continues to cast its shadow as I discussed earlier.

Meanwhile political extremism, religious intolerance and the return to authoritarian rule in the parts of the Arab world are increasing tensions in the whole region. This places both the civilian populations and the governing regimes under intense pressure at home and across borders. If, as the laws of physics tell us, the pressure will inevitably breach its limit and cause an explosion. And Those who do not learn from history are doomed to repeat it.

But the future may not be as bright as the propaganda portrays. With the Arab world still not fully embracing the data economy, while many of its population and businesses not fully appreciating the new reality that data is the new oil. The lack of investing in research and development and activity in the Arab world, will have adverse effects on its economic and technological progress. Nevertheless, with the internet becoming ubiquitous in the Arab world, people are becoming more aware and there are green shots of change. There is hope, but there are risks. The future is Amber!

## 6.2    Signposts and Questions

In the last section of Chapter 1, I listed several events to watch out for, which I predict will shape the Arab world's future in the coming years and which will have global implications. Readers can refer to Section 1.7 for the full list.

In a nutshell the main events to watch for are the polarisation in the Middle East and globally igniting new conflicts; the continuing Cold War between the USA, China and Russia; major civil unrest in the USA; the volatility of the oil price; deglobalisation, energy transition and climate change policies.

Taking all of the above points into consideration, I would like to pose few questions about the future:

- Where is the oil and natural gas industry heading to? Has oil price bottomed out? Will capital spending shift from fossil fuels to alternatives? Will there be further gains in productivity, efficiency or energy technology? This line of questions continues to be relevant, but the prospects for oil and natural gas are cloudier.

- Will humanity embrace the hydrogen age and witness a new Energy Age where alternatives dominate? Or is that a false hope?

- Will CCS be widely adopted as an interim solution to curb emissions? Or is it just a public relations exercise, (especially if we take into account the unfeasibility of its implementation)?

- Is there really an optimal point for competing fuel prices? Which fuel will win? Which fuel will fulfil the Trilemma (affordable, available, clean) to be successful? Does natural gas stand a chance?

- With tensions boiling in the Arab world, will there be another Arab Spring in the next five years? And if yes will it succeed? Will there continue to be instability, conflict and destruction?

- Will China invade Taiwan in the next five years? How will the West respond? What will the implications be on the energy and the industrial age when so many industries are reliant on the Taiwanese produced chips and semiconductors?

- With democracy in evident retreat, will deglobalisation and nationalism lead us further along the path towards a volatile unstable world?

- What are the implications of the erosion of privacy and restrictions on civil liberties worldwide? Will governments that used the pandemic as a pretext for greater surveillance on their populations, willingly roll it back? Or will people accept the erosion of their civil liberties, as they did after September 11th?

Events in the last few years confirmed that we are indeed live in a post-truth world. The melodrama of the latest American elections and the unfounded lies by Trump and his administration illustrated this fact in front of all our eyes. There are genuine fears that Trump era will be America's Gorbachev's Soviet era, whose actions led to the disintegration on the nation. In the UK Boris Johnson could be the British instigator of the disintegration of the United Kingdom, leading to a United Ireland and the eventual collapse of the EU, with various political groups advocating exiting the EU in Sweden, Netherlands, France and Italy.

This failure of the democratic model is music to Putin's and Chinese ears. China has confidently demonstrated to its citizens, and the world, that its governance model is superior. Its handling of the Covid-19 pandemic, compared to the Western world and other adversaries (e.g. India), appear to vindicate its propaganda. The pandemic has been an accelerant to the pressures on Western nations that may lead to Asia's pre-eminence. It appears that this is Asia's century,[6] China is now the challenger. It was challenged by the West previously to adjust its economic model, but it is now proving to be on the ascendant and its state unfree economy is throwing down the gauntlet to the West.

The big challenge to all of us in the next few years is confronting misinformation. This will be vital. What happened with Covid-19, spreading of fake news, the deliberate spreading of lies, half-truths, misinformation and conspiracy theories is a wakeup call.

We are living in an unprecedented, transformational period, with AI, machine learning, big data and data analytics becoming the new tools that are shaping our lives. The new attitudes towards working from home, transportation modes, public health and national security are rapidly reshaping the world. The Covid-19 pandemic accelerated many of these changes and its impacts will continue for years. A journey that involves accelerated innovation has just been kick-started.

In the next five years things will inevitably continue to change,

---

[6] https://www.bloomberg.com/graphics/2020-global-economic-forecast-2050/

what we forecast today will in due course, become the past. We need to remember that every event is news before it is history. Some of the above questions will be answered, while others may be deferred. While some nations will prosper, others will decline. It is a continuous cycle. Where does the future hold for Arabs? They stand a chance if they embrace change, invest in research and development, encourage social and institutional development, instead of unnecessary mega projects. For many young people there, they not only need to believe that there is hope, but they need to work for it, their goals will not happen without their action.

The Arab World could excel as a leader, by realising it has the opportunity in the next few years to capitalise on its oil and gas wealth to become a leader in alternative energy; building on the strengths; utilising its vast renewable energy resources; investing in its human capital, enhancing education to nurture localised research and development culture. The latter requires a significant shift in mentality, where emphasis should be on promoting free thinking and innovation, rather than tradition or indoctrination.

The Arab decision makers must rectify the conditions that harm the population, leaving it ripe for recruitment by fundamentalists or dictators who will reward them richly in exchange for escape from poverty. Knowledge is power so decision makers need to empower the population, releasing them from the fatalism that has held them back for generations. If they let go of grudges and feuds that harm everyone, instead striving to raise each other up – tribes, communities, nations, humanity – then they will thrive. This may sound revolutionary, but they have done it before and, if they do it again, they may not just catch the boat but upgrade it and operate the helm.

# APPENDICES

**Appendix I** : **Arab World Statistics**

**Appendix II** : **Oil and Natural Gas Data**[1]

---

[1] In all graphs in this appendix the area referred to as world excludes the Arab world, hence the summation gives the overall numerical quantity.

# Appendix I: Arab World Statistics

*Table A1.1: Arab world's basic information*

| Country | Capital | Area km² | Rank | Population million | Rank |
|---------|---------|----------|------|--------------------|------|
| Bahrain | Manama | 760 | 22 | 1.527 | 20 |
| Iraq | Baghdad | 438317 | 10 | 39.650 | 4 |
| Jordan | Amman | 89342 | 14 | 10.910 | 11 |
| Kuwait | Kuwait City | 17818 | 17 | 3.032 | 18 |
| Lebanon | Beirut | 10400 | 19 | 5.261 | 14 |
| Oman | Muscat | 309500 | 11 | 3.695 | 17 |
| Palestine | East Jerusalem | 6220 | 20 | 4.906 | 15 |
| Qatar | Doha | 11586 | 18 | 2.480 | 19 |
| Saudi Arabia | Riyadh | 2149690 | 2 | 34.784 | 6 |
| Syria | Damascus | 187437 | 12 | 20.384 | 8 |
| UAE | Abu Dhabi | 83600 | 15 | 9.857 | 12 |
| Yemen | Sana'a | 527968 | 9 | 30.399 | 7 |
| Algeria | Algiers | 2381740 | 1 | 43.577 | 3 |
| Comoros | Moroni | 2235 | 21 | 0.864 | 22 |
| Djibouti | Djibouti | 23200 | 16 | 0.938 | 21 |
| Egypt | Cairo | 1001450 | 6 | 106.437 | 1 |
| Libya | Tripoli | 1759540 | 4 | 7.017 | 13 |
| Mauritania | Nouakchott | 1030700 | 5 | 4.079 | 16 |
| Morocco | Rabat | 716550 | 7 | 36.562 | 5 |
| Somalia | Mogadishu | 637657 | 8 | 12.095 | 9 |
| Sudan | Khartoum | 1861484 | 3 | 46.751 | 2 |
| Tunisia | Tunis | 163610 | 13 | 11.811 | 10 |
| *ARAB WORLD* | | 13410804 | | 437.017 | |
| *Arab world Share (%)* | | 8.91 | | 5.62 | |

Source: CIA Factbook.
Note 1: Population as estimated in 2021.
Note 2: Area and population of Morocco include Western Sahara
Note 3: Area and population of Palestine is for the West Bank including East Jerusalem and Gaza Strip. Population number excludes Israeli settlers.
Note 4: East Jerusalem is the proclaimed Palestinian capital. Ramallah is the current administrative centre of the Palestinian Authority, while Gaza city is the administrative centre of the Hamas-led government.

# Appendix II: Oil and Natural Gas Data

## *Oil Reserves*

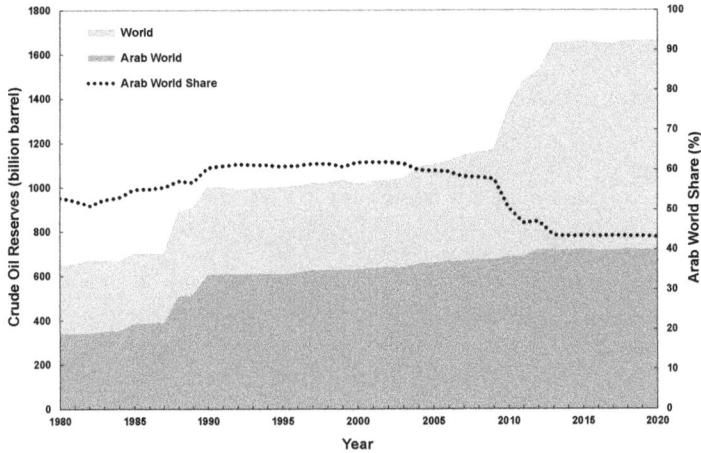

*Figure A2.1: Arab world conventional proved oil reserves and its share to the world's total (1980-2020)*

Source: EIA (http://www.eia.doe.gov)

*Table A2.1: World conventional proved oil reserves – top ten countries (2020)*

| Rank | Country | Crude Oil Reserves | Share |
|------|---------|-----|-----|
| | | billion barrel | % |
| 1 | Venezuela | 302.81 | 18.22 |
| 2 | Saudi Arabia | 267.03 | 16.07 |
| 3 | Iran | 155.60 | 9.36 |
| 4 | Iraq | 145.02 | 8.73 |
| 5 | Kuwait | 101.50 | 6.11 |
| 6 | United Arab Emirates | 97.80 | 5.88 |
| 7 | Russia | 80.00 | 4.81 |
| 8 | Libya | 48.36 | 2.91 |
| 9 | United States | 47.11 | 2.83 |
| 10 | Nigeria | 36.97 | 2.22 |
| | *TOTAL TOP TEN* | 1282.20 | 77.15 |
| | *WORLD* | 1661.91 | |

Source: EIA (http://www.eia.doe.gov)
Note 1: Totals may not add up due to rounding.

## *Oil Reserves*

*Table A2.2: Conventional proved oil reserves in the Arab countries (2020)*

| Country | Crude Oil Reserves | Rank | Share |
|---|---|---|---|
| | billion barrel | | % |
| Bahrain | 0.09 | 14 | 0.01 |
| Iraq | 145.02 | 2 | 8.73 |
| Jordan | 0.00 | 16 | 0.00 |
| Kuwait | 101.50 | 3 | 6.11 |
| Lebanon | 0.00 | | |
| Oman | 5.37 | 8 | 0.32 |
| Palestine | 0.00 | | |
| Qatar | 25.24 | 6 | 1.52 |
| Saudi Arabia | 267.03 | 1 | 16.07 |
| Syria | 2.50 | 12 | 0.15 |
| United Arab Emirates | 97.80 | 4 | 5.88 |
| Yemen | 3.00 | 11 | 0.18 |
| Algeria | 12.20 | 7 | 0.73 |
| Comoros | 0.00 | | |
| Djibouti | 0.00 | | |
| Egypt | 3.30 | 10 | 0.20 |
| Libya | 48.36 | 5 | 2.91 |
| Mauritania | 0.02 | 15 | 0.00 |
| Morocco | 0.00 | 17 | 0.00 |
| Somalia | 0.00 | | |
| Sudan | 5.00 | 9 | 0.30 |
| Tunisia | 0.43 | 13 | 0.03 |
| *ARAB WORLD* | 716.87 | | 43.14 |

Source: EIA (http://www.eia.doe.gov)
Note 1: Totals may not add up due to rounding.
Note 2: Number for Sudan includes South Sudan.

## Oil Production[1]

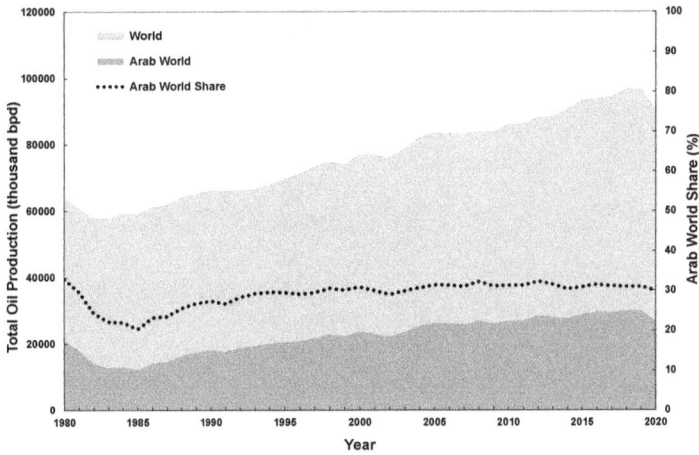

*Figure A2.2: Arab world total oil production and its share to the world's total (1980-2020)*

Source: EIA (http://www.eia.doe.gov)

*Table A2.3: World total oil production – top ten countries (2020)*

| Rank | Country | Total Oil Production | Share |
|---|---|---|---|
| | | thousand bpd | % |
| 1 | United States | 17405.01 | 19.28 |
| 2 | Saudi Arabia | 10740.36 | 11.90 |
| 3 | Russia | 10503.87 | 11.64 |
| 4 | Canada | 5215.01 | 5.78 |
| 5 | China | 4224.51 | 4.68 |
| 6 | Iraq | 4159.15 | 4.61 |
| 7 | United Arab Emirates | 3785.20 | 4.19 |
| 8 | Brazil | 3112.46 | 3.45 |
| 9 | Iran | 3014.48 | 3.34 |
| 10 | Kuwait | 2746.05 | 3.04 |
| | *TOTAL TOP TEN* | 64906.09 | 71.92 |
| | *WORLD* | 90252.77 | |

Source: EIA (http://www.eia.doe.gov)
Note 1: Totals may not add up due to rounding.
## Oil Production

[1] Total oil production includes crude oil, condensate, natural gas liquids and refinery gain, but excludes other liquids

*Table A2.4: Total oil production in the Arab countries (2020)*

| Country | Total Oil Production | Rank | Share |
|---|---|---|---|
| | thousand bpd | | % |
| Bahrain | 185.38 | 10 | 0.21 |
| Iraq | 4159.15 | 2 | 4.61 |
| Jordan | 0.40 | 15 | 0.00 |
| Kuwait | 2746.05 | 4 | 3.04 |
| Lebanon | 0.00 | | |
| Oman | 958.58 | 7 | 1.06 |
| Palestine | 0.00 | | |
| Qatar | 1733.37 | 5 | 1.92 |
| Saudi Arabia | 10740.36 | 1 | 11.90 |
| Syria | 42.94 | 13 | 0.05 |
| United Arab Emirates | 3785.20 | 3 | 4.19 |
| Yemen | 66.29 | 12 | 0.07 |
| Algeria | 1404.65 | 6 | 1.56 |
| Comoros | 0.00 | | |
| Djibouti | 0.00 | | |
| Egypt | 602.70 | 8 | 0.67 |
| Libya | 416.78 | 9 | 0.46 |
| Mauritania | 0.00 | | |
| Morocco | 0.05 | ) | |
| Somalia | 0.00 | | |
| Sudan | 66.93 | 11 | 0.07 |
| Tunisia | 34.71 | 14 | 0.04 |
| *ARAB WORLD* | 26943.55 | | 29.85 |

Source: EIA (http://www.eia.doe.gov)
Note 1: Totals may not add up due to rounding.
Note 2: Number for Sudan includes South Sudan.

## Natural Gas Reserves

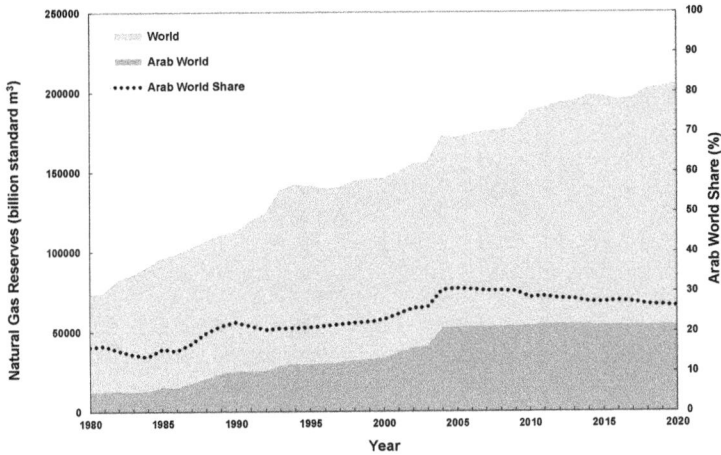

*Figure A2.3: Arab world conventional proved Natural gas and its share to the world's total (1980-2020)*

Source: EIA (http://www.eia.doe.gov)

*Table A2.5: World conventional proved natural gas reserves – top ten countries (2020)*

| Rank | Country | Natural Gas Reserves | Share |
|------|---------|---------------------|-------|
|      |         | billion standard m³ | % |
| 1 | Russia | 47804.84 | 23.26 |
| 2 | Iran | 33898.03 | 16.50 |
| 3 | Qatar | 23860.31 | 11.61 |
| 4 | United States | 13178.68 | 6.41 |
| 5 | Turkmenistan | 9910.80 | 4.82 |
| 6 | Saudi Arabia | 9068.58 | 4.41 |
| 7 | China | 6314.06 | 3.07 |
| 8 | United Arab Emirates | 6090.84 | 2.96 |
| 9 | Nigeria | 5674.84 | 2.76 |
| 10 | Venezuela | 5673.85 | 2.76 |
|   | *TOTAL TOP TEN* | 161474.84 | 78.58 |
|   | *WORLD* | 205497.98 | |

Source: EIA (http://www.eia.doe.gov)
Note 1: Totals may not add up due to rounding.

### Natural Gas Reserves

*Table A2.6: Conventional proved natural gas reserves in the Arab countries (2020)*

| Country | Natural Gas Reserves | Rank | Share |
|---|---|---|---|
| | billion standard m$^3$ | | % |
| Bahrain | 192.55 | 12 | 0.09 |
| Iraq | 3728.90 | 5 | 1.81 |
| Jordan | 6.03 | 16 | 0.00 |
| Kuwait | 1783.94 | 6 | 0.87 |
| Lebanon | 0.00 | | |
| Oman | 651.28 | 9 | 0.32 |
| Palestine | 0.00 | | |
| Qatar | 23860.31 | 1 | 11.61 |
| Saudi Arabia | 9068.58 | 2 | 4.41 |
| Syria | 240.69 | 11 | 0.12 |
| United Arab Emirates | 6090.84 | 3 | 2.96 |
| Yemen | 478.55 | 10 | 0.23 |
| Algeria | 4503.87 | 4 | 2.19 |
| Comoros | 0.00 | | |
| Djibouti | 0.00 | | |
| Egypt | 1783.94 | 6 | 0.87 |
| Libya | 1504.86 | 8 | 0.73 |
| Mauritania | 28.32 | 15 | 0.01 |
| Morocco | 1.44 | 18 | 0.00 |
| Somalia | 5.66 | 17 | 0.00 |
| Sudan | 84.95 | 13 | 0.04 |
| Tunisia | 65.13 | 14 | 0.03 |
| *ARAB WORLD* | 54079.85 | | 26.32 |

Source: EIA (http://www.eia.doe.gov)
Note 1: Totals may not add up due to rounding.
Note 2: Number for Sudan includes South Sudan.

## Natural Gas Production[2]

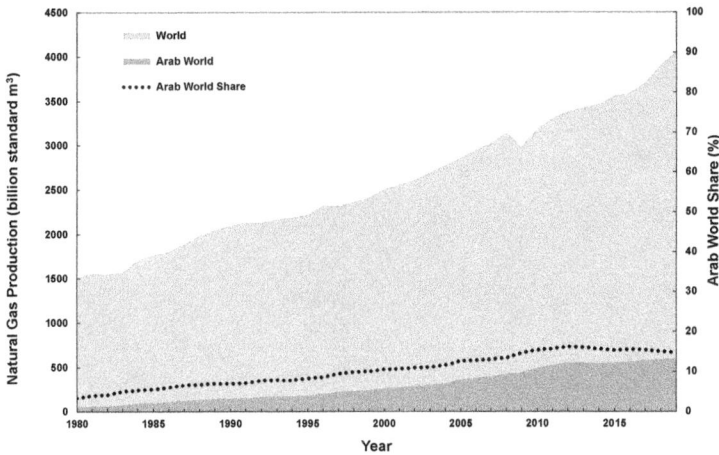

*Figure A2.4: Arab world total dry natural gas production and its share to the world's total (1980-2019)*

Source: EIA (http://www.eia.doe.gov)

*Table A2.7: World total dry natural gas production – top ten countries (2019)*

| Rank | Country | Dry Natural Gas Production | Share |
|------|---------|---------------------------|-------|
| | | billion standard m³ | % |
| 1 | United States | 961.85 | 23.81 |
| 2 | Russia | 677.83 | 16.78 |
| 3 | Iran | 237.56 | 5.88 |
| 4 | China | 179.32 | 4.44 |
| 5 | Canada | 178.72 | 4.42 |
| 6 | Qatar | 167.46 | 4.14 |
| 7 | Australia | 145.95 | 3.61 |
| 8 | Norway | 114.88 | 2.84 |
| 9 | Saudi Arabia | 112.76 | 2.79 |
| 10 | Algeria | 87.85 | 2.17 |
| | *TOTAL TOP TEN* | 2864.18 | 70.89 |
| | *WORLD* | 4040.42 | |

Source: EIA (http://www.eia.doe.gov)
Note 1: Totals may not add up due to rounding.

## Natural Gas Production

---

[2] Natural gas production refers to dry production

*Table A2.8: Total dry natural gas production in the Arab countries (2019)*

| Country | Dry Natural Gas Production | Rank | Share |
|---|---|---|---|
| | billion standard m$^3$ | | % |
| Bahrain | 16.90 | 7 | 0.42 |
| Iraq | 0.91 | 12 | 0.02 |
| Jordan | 0.20 | 13 | 0.00 |
| Kuwait | 15.04 | 8 | 0.37 |
| Lebanon | 0.00 | | |
| Oman | 30.89 | 6 | 0.76 |
| Palestine | 0.00 | | |
| Qatar | 167.46 | 1 | 4.14 |
| Saudi Arabia | 112.76 | 2 | 2.79 |
| Syria | 3.53 | 10 | 0.09 |
| United Arab Emirates | 62.89 | 5 | 1.56 |
| Yemen | 0.09 | 15 | 0.00 |
| Algeria | 87.85 | 3 | 2.17 |
| Comoros | 0.00 | | |
| Djibouti | 0.00 | | |
| Egypt | 64.29 | 4 | 1.59 |
| Libya | 10.56 | 9 | 0.26 |
| Mauritania | 0.00 | | |
| Morocco | 0.11 | 14 | 0.00 |
| Somalia | 0.00 | | |
| Sudan | 0.00 | | |
| Tunisia | 1.03 | 11 | 0.03 |
| ARAB WORLD | 574.51 | | 14.22 |

Source: EIA (http://www.eia.doe.gov)
Note 1: Totals may not add up due to rounding.
Note 2: Number for Sudan includes South Sudan.